GOVERNMENT PROCEDURES AND OPERATIONS

TECHNOLOGY POLICIES AND ISSUES FOR THE U.S. GOVERNMENT

GOVERNMENT PROCEDURES AND OPERATIONS

Additional books in this series can be found on Nova's website
under the Series tab.

Additional E-books in this series can be found on Nova's website
under the E-book tab.

GOVERNMENT PROCEDURES AND OPERATIONS

TECHNOLOGY POLICIES AND ISSUES FOR THE U.S. GOVERNMENT

DANIELLE A. GODDEN
AND
MASON C. ANDREWS
EDITOR

Nova Science Publishers, Inc.
New York

For permission to use material from this book please contact us:
Telephone 631-231-7269; Fax 631-231-8175
Web Site: http://www.novapublishers.com

Library of Congress Cataloging-in-Publication Data

Technology policies and issues for the U.S. government / editors, Danielle A. Godden and Mason C. Andrews.

p. cm.

Includes index.

ISBN 978-1-61209-501-1 (hardcover)

1. Technology and state--United States. 2. Research, Industrial--Government policy--United States. 3. Biotechnology--Research--Government policy--United States. I. Godden, Danielle A. II. Andrews, Mason C.

T21.T437 2011

338.973--dc22

2011003594

Published by Nova Science Publishers, Inc. † New York

CONTENTS

PREFACE

This new book examines technology policies and issues for the U.S. Government. Topics discussed include the appointment of a federal chief technology officer (CTO) during the Obama administration; U.S. initiatives to promote global internet freedom; biotechnology in animal agriculture; industrial competitiveness and technological advancement; the Federal Networking and Information Technology Research and Development program, as well as an overview of the National Nanotechnology Initiative.

Chapter 1- In November 2007, Senator Barack Obama announced his intention, if elected President, to appoint a federal chief technology officer (CTO). He also identified several specific areas of responsibility of the CTO including transparency of government operations, computer and network security (sometimes referred to as cybersecurity), identification and adoption of best technologies and practices by federal agencies, and interoperability of emergency communications technologies for first responders.

Chapter 2- Modern means of communications, led by the Internet, provide a relatively inexpensive, open, easy-entry means of sharing ideas, information, pictures, and text around the world. In a political and human rights context, in closed societies when the more established, formal news media is denied access to or does not report on specified news events, the Internet has become an alternative source of media, and sometimes a means to organize politically.

Chapter 3- Animal agriculture is being transformed by rapid advances in biotechnology—a term that encompasses a variety of technologies, including genetic engineering (GE), genetic modification, transgenics, recombinant DNA techniques, and cloning, among others. Producers are interested in the application of biotechnology to improve productivity, consistency, and quality; to introduce new food, fiber, and medical products; and to protect the environment. Potential human health applications of transgenic animals include producing biopharmaceuticals and generating organs, tissues, and cells for xenotransplantation. Criticisms of such applications involve issues ranging from food safety and social resistance to potential negative impacts on animal welfare and on ecosystems. Questions also have arisen about the adequacy of the current regulatory structure to assess and manage any risks created by these technologies.

Chapter 4- U.S. soybean, cotton, and corn farmers have rapidly adopted genetically engineered (GE) varieties of these crops since their commercialization in the mid-1990s. Over the last decade, GE varieties in the United States have increased from 3.6 million acres to 143 million acres. Worldwide, 25 countries planted GE crops on approximately 309 million acres in 2008. GE varieties now dominate soybean, cotton, and corn production in the United States, and they continue to expand rapidly in other countries. As adoption has spread, policy debates have continued over the costs and benefits of GE products.

Chapter 5- In May 2003, the United States, Canada, and Argentina initiated a dispute with the European Union concerning the EU's *de facto* moratorium on biotechnology product approvals, in place since 1998. Although the EU effectively lifted the moratorium in May 2004 by approving a genetically engineered (GE) corn variety (MON810), the three complainants pursued the case, in part because a number of EU member states continue to block already approved biotech products. Industry estimates are that the moratorium costs U.S. corn growers some $300 million in exports to the EU annually. Corn gluten exports from the United States to the EU have been blocked since 2007 because of a zero tolerance policy governing the accidental presence of non-approved U.S. GE corn in such shipments.

Chapter 6- There is ongoing interest in the pace of U.S. technological advancement due to its influence on U.S. economic growth, productivity, and international competitiveness. Because technology can contribute to economic growth and productivity increases, congressional attention has focused on how to augment private-sector technological development. Legislative activity over the past 25 or more years has created a policy for technology development, albeit an ad hoc one. Because of the lack of consensus on the scope and direction of a national policy, Congress has taken an incremental approach aimed at creating new mechanisms to facilitate technological advancement in particular areas and making changes and improvements as necessary.

Chapter 7- In the early 1990s, Congress recognized that several federal agencies had ongoing high-performance computing programs, but no central coordinating body existed to ensure long-term coordination and planning. To provide such a framework, Congress passed the High-Performance Computing and Communications Program Act of 1991 (P.L. 102-194) to enhance the effectiveness of the various programs. In conjunction with the passage of the act, the White House Office of Science and Technology Policy (OSTP) released *Grand Challenges: High-Performance Computing and Communications*. That document outlined a research and development (R&D) strategy for high-performance computing and a framework for a multiagency program, the High-Performance Computing and Communications (HPCC) Program. The HPCC Program has evolved over time and is now called the Networking and Information Technology Research and Development (NITRD) Program, to better reflect its expanded mission.

Chapter 8- Nanotechnology—a term encompassing the science, engineering, and applications of submicron materials—involves the harnessing of unique physical, chemical, and biological properties of nanoscale substances in fundamentally new and useful ways. The economic and societal promise of nanotechnology has led to substantial and sustained investments by governments and companies around the world. In 2000, the United States launched the world's first national nanotechnology program. From FY2001 through FY2010, the federal government invested approximately $12.4 billion in nanoscale science, engineering, and technology through the U.S. National Nanotechnology Initiative (NNI). U.S. companies and state governments have invested billions more. President Obama has requested an additional $1.8 billion in NNI funding for FY2011. As a result of this focus and these investments, the United States has, in the view of many experts, emerged as a global leader in nanotechnology. However, the competition for global leadership in nanotechnology is intensifying as countries and companies around the world increase their investments.

In: Technology Policies and Issues for the U.S. Government ISBN: 978-1-61209-501-1
Editors: Danielle A. Godden and Mason C. Andrews

Chapter 1

A FEDERAL CHIEF TECHNOLOGY OFFICER IN THE OBAMA ADMINSTRATION: OPTIONS AND ISSUES FOR CONSIDERATION*

John F. Sargent, Jr.

SUMMARY

In November 2007, Senator Barack Obama announced his intention, if elected President, to appoint a federal chief technology officer (CTO). He also identified several specific areas of responsibility of the CTO including transparency of government operations, computer and network security (sometimes referred to as cybersecurity), identification and adoption of best technologies and practices by federal agencies, and interoperability of emergency communications technologies for first responders.

On April 18, 2009, President Obama appointed Virginia Secretary of Technology Aneesh P. Chopra to serve in the newly created position of federal chief technology officer. In announcing the appointment, the President indicated that Mr. Chopra would undertake roles beyond what might be considered traditional CTO responsibilities. As the President described them, these roles include promoting technological innovation to help the United States create jobs, reduce health care costs, protect the homeland, and address other national goals. Mr. Chopra serves as assistant to the President and chief technology officer, as well as associate director for technology in the White House Office of Science and Technology Policy. Under a provision of Executive Order 13500, issued on February 5, 2009, he also serves as a member of the White House Domestic Policy Council in his capacity as assistant to the President and CTO.

The CTO may face a variety of challenges in executing the mission envisioned by the President. Among the early challenges will be negotiating domains of responsibilities within

---

* This is an edited, reformatted and augmented version of Congressional Research Service publication, Report R40150, dated June 4, 2010.

the White House and with executive branch agencies that have overlapping missions. Some commentators have expressed concerns about the impact the creation of a CTO might have on existing offices and agencies with respect to the allocation and coordination of authorities and responsibilities. Other commentators have asserted that a high-level CTO could serve as an advocate for technological innovation and foster increased knowledge sharing among federal agencies to more effectively implement information technology solutions to meet disparate mission requirements. Mr. Chopra's appointment as both CTO and associate director for technology at OSTP may address, in part, questions related to mission alignment, coordination, and integration. Since assuming his dual roles, Mr. Chopra has publicly engaged in discussions covering a wide range of technology policy-related areas, including research and development, innovation, open government, government performance, education, science and engineering workforce, health care information technology, broadband, patent reform, and net neutrality.

Congress faces President Obama's appointment of Mr. Chopra and the President's stated plans for the federal CTO. There is currently no formal position description for the CTO. Accordingly, the official duties of the CTO remain largely undefined. Congress may elect to provide a statutory foundation for the CTO, define the roles and authorities of the CTO, authorize and appropriate funds, provide for oversight, and address other aspects of the position.

Overview

In November 2007, Senator Barack Obama announced his intention, if elected President, to appoint a federal chief technology officer (CTO). On April 18, 2009, President Obama appointed Virginia Secretary of Technology Aneesh P. Chopra to serve as "America's Chief Technology Officer."[1]

Mr. Chopra serves as assistant to the President and CTO, and holds the position of associate director for technology in the White House Office of Science and Technology Policy.[2] In his capacity as assistant to the President and CTO, Mr. Chopra also serves as a member of the White House Domestic Policy Council under a provision of Executive Order 13500 (amending Executive Order 12859) issued by President Obama on February 5, 2009.[3] In announcing the appointment of Mr. Chopra, President Obama broadly defined the role of the CTO as promoting "technological innovation to help achieve our most important priorities—from creating jobs and reducing health care costs to keeping our nation secure."[4]

This report presents President Obama's vision for the CTO position put forth during his campaign for the presidency and during the presidential transition. It also discusses the potential scope of duties and authorities of the CTO, articulates organizational precedents and challenges for the CTO, reviews some of the activities undertaken by Aneesh Chopra during his tenure as CTO, and identifies issues Congress may choose to consider if it opts to exert oversight or to develop legislation to create the position and/or office of the CTO.

BACKGROUND

In 2008, Barack Obama's presidential campaign issued a position paper on the issue of technology and innovation which included the following description of the role envisioned for a CTO:

> Bring Government into the 21st Century: Barack Obama will use technology to reform government and improve the exchange of information between the federal government and citizens while ensuring the security of our networks. Obama believes in the American people and in their intelligence, expertise, and ability and willingness to give and to give back to make government work better.
>
> Obama will appoint the nation's first Chief Technology Officer (CTO) to ensure that our government and all its agencies have the right infrastructure, policies and services for the 21st century. The CTO will ensure the safety of our networks and will lead an interagency effort, working with chief technology and chief information officers of each of the federal agencies, to ensure that they use best-in-class technologies and share best practices.
>
> The CTO will have a specific focus on transparency, by ensuring that each arm of the federal government makes its records open and accessible as the E-Government Act requires. The CTO will also focus on using new technologies to solicit and receive information back from citizens to improve the functioning of democratic government.
>
> The CTO will also ensure technological interoperability of key government functions. For example, the Chief Technology Officer will oversee the development of a national, interoperable wireless network for local, state and federal first responders as the 9/11 commission recommended. This will ensure that fire officials, police officers and [emergency medical technicians] from different jurisdictions have the ability to communicate with each other during a crisis and we do not have a repeat of the failure to deliver critical public services that occurred in the aftermath of Hurricane Katrina.[5]

In November 2008, President-elect Obama reiterated his intentions for the CTO on Change.gov, the website of the Office of the President-elect:

> Bring Government into the 21st Century: Use technology to reform government and improve the exchange of information between the federal government and citizens while ensuring the security of our networks. Appoint the nation's first Chief Technology Officer (CTO) to ensure the safety of our networks and lead an interagency effort, working with chief technology and chief information officers of each of the federal agencies, to ensure that they use best-in-class technologies and share best practices.[6]

Prior to Mr. Chopra's appointment, details related to the CTO position remained uncertain. There was little or no information on the specific duties and authorities of the CTO and how its functions will be coordinated and integrated within OSTP, with other White House offices, and with federal departments and agencies. During this period, some commentators speculated about the range of duties and authorities the CTO may be given (see discussion on pages 3-5 of this report). While many details remain uncertain about the CTOs role, some aspects have become more clear as Mr. Chopra has undertaken initiatives, testified before Congress, and given interviews (see "CTO Structure and Activities in the Obama Administration").

Potential Scope of Duties and Authorities

A fundamental question for policymakers related to the CTO position is: What should be the scope of duties and authorities of this position? Two broad roles for the CTO appear to have emerged.

The first role, explicitly articulated on President Obama's campaign and transition websites, might be described as a supra-chief information officer with the mission of using information technology to improve the delivery of government services, increasing transparency of government policymaking, and opening channels for increased citizen participation in government, as well as ensuring that the nation's information and communications infrastructure is robust and secure. In this capacity, the CTO would also ensure that "best practices" are identified, shared, and implemented across agencies.

The second role for the CTO might be described as an advocate for technological innovation in support of national interests such as economic growth, job creation, improvements to quality of life, national defense, and homeland security. President Obama's dual appointment of Mr. Chopra to serve as both CTO and OSTP associate director for technology, and his statement accompanying the selection, positioned Mr. Chopra to undertake such a role.

Prior to Mr. Chopra's appointment, several analysts proposed a variety of roles for an Obama administration CTO. Stanford University professor Lawrence Lessig asserted that the CTO could play an important role in bringing an understanding of how technology might be used to address national challenges:

> The CTO could be a critically important position, from deciding how to make government more efficient and transparent through technology, to helping advance public policy questions like those surrounding global warming.[7]

The CTO could play an important role in national economic policy according to Andrew D. Lipman, a telecommunications attorney with Bingham McCutchen, LLP. Mr. Lipman asserted that

> [President] Obama sees greater broadband penetration as an enormous economic engine, much like the railroads were a century ago. That is why the CTO will play such a critical role in any recovery plan.[8]

Ed Felten, a Princeton University professor of computer science and public affairs and director of Princeton's Center for Information Technology Policy, likewise suggested a broad role for the CTO in innovation policy:

> The CTO could act as the cybersecurity czar, ensuring that reliability of the government infrastructure is protected. And much like the role of the presidential science advisor, the CTO could offer advice to the President on all areas of technology. The role could be a catalyst to push us closer to being a more entrepreneurial, high-tech country.[9]

The Center for American Progress (CAP) and New Democracy Project (NDP), public policy institutes, proposed that the CTO serve as a champion for information and communications technologies broadly. CAP/NDP proposed that the CTO be given three roles: advising the President on the use of ICT to create a more open and efficient government; working with OSTP to advise the President on all issues that have a critical ICT component, including economic, national security, health care, and education policies; and helping to ensure ICT platforms deployed across the United States are robust, broadly available, and affordable to all sectors. CAP/NDP anticipate this latter role would "straddle the public and private sectors" to maximize private innovation for the public good.[10]

In contrast, the Association of American Universities (AAU), an association of research universities, recommended to President Obama that a CTO be given a more narrow role:

> If a new Chief Technology Officer position is created, this individual's responsibilities primarily should be to oversee the use of technology within the White House and to coordinate the use of technology within the Executive Branch.[11]

In addition, the AAU urged that OSTP remain the sole presidential advisor for science and technology:

> [The president should] retain a single presidential advisor for science and technology with a strong associate director for technology and innovation.... The responsibility for coordinating and helping to shape science and technology policy across government agencies should remain that of the OSTP Director, as has historically been the case.[12]

Three factors are likely to play key roles in defining the full scope of duties and authorities of the CTO: first, the role that President Obama wants the CTO to play (including authorities derived from formal means, such as an executive order, and non-formal means, such as presidential expressions of support for, and confidence in, the CTO); second, congressional oversight of any actions by the Obama Administration; and, third, statutory duties and authorities, if any, that Congress may elect to confer upon the position and/or office.

Further, the personal attributes of the CTO may, in part, define the role of CTO. In this regard, Mr. Chopra served as Virginia's Secretary of Technology where he led "the Commonwealth's strategy to effectively leverage technology in government reform, to promote Virginia's innovation agenda, and to foster technology-related economic development."[13] Previously, Chopra served as managing director of the Advisory Board Company, a public company that provides best practices research and analysis to the health care industry. He led the firm's Financial Leadership Council and the Working Council for Health Plan Executives.[14]

Organizational Precedents

In defining or refining the duties and authorities of the CTO, Congress and the Obama Administration may opt to draw from the duties and authorities of existing and previous organizational structures and positions. These could include agency-level chief information officers (CIOs) and CTOs; the Bush Administration's use of the Office of Management and Budget (OMB) to handle CIO, CTO, and e-government responsibilities; the Clinton

Administration's "reinventing government" initiative; the Commerce Department's recently eliminated Technology Administration and its agency predecessors; and the White House Office of Science and Technology Policy (OSTP). An overview of each is provided below.

Agency CIOs, CTOs, and the Chief Information Officers Council (CIO Council)

As information technology began to play an increasingly important (and more costly) role in federal agency operations and services, agencies began to appoint CIOs to manage information technology systems and acquisitions. According to the CIO Council (discussed below) the mandate of federal CIOs is

> to ensure the rapid and effective implementation of information management and information technology (IM/IT) solutions within each agency and to create a more results-oriented, efficient, and citizen-centered Federal government. [15]

In 1996, Congress enacted the Clinger-Cohen Act[16] which, among other things, requires the establishment of a CIO in each federal agency. The duties assigned CIOs under the act include providing information management advice and policy to the agency head; developing, maintaining, and facilitating information systems; and evaluating, assessing, and reporting to the agency head on the progress made developing agency information technology systems.[17]

Following passage of the Clinger-Cohen Act, President Bill Clinton issued Executive Order 13011[18] which directs agencies to establish the position of chief information officer to provide clear accountability for information resources management activities; identifies the broad responsibilities of federal agency CIOs; and establishes a federal Chief Information Officer Council (CIO Council) chaired by the OMB deputy director for management. The CIO Council was later codified by the E-Government Act of 2002 (P.L. 107-347), which designated it

> the principal interagency forum for improving agency practices related to the design, acquisition, development, modernization, use, operation, sharing, and performance of Federal Government information resources.[19]

No parallel council exists in the federal government for CTOs, nor has Congress statutorily defined the duties, responsibilities, and qualifications of CTOs in a manner comparable to that for CIOs under the Clinger-Cohen Act. The position of chief technology officer first emerged in the private sector in the 1980s. Since then, some federal agencies have established CTO positions. In general, a CTO is responsible for monitoring, assessing, and selecting new technologies for applications to improve an organization's performance. Such applications of technology can be focused on developing new products and services or on improving internal processes. Given the ubiquitous role technology plays in most organizations today, CTOs often work with a wide variety of stakeholders, including chief executives, scientists, engineers, research managers, and marketing professionals.[20] Generally, federal agency CTOs are responsible for developing and/or implementing technologies that enable the agency to better perform its missions. The role of a CTO in a federal agency can be similar to, and the title sometimes used interchangeably with, the CIO.[21] In many cases, agency CTOs report to the CIO and have a focus on information technology. However, some CTOs have responsibility for a wider scope of technologies and technical issues. For example, the CTO for the Department of Homeland Security's Transportation Security Administration is responsible for "implementation and development of security technologies across several modes of transportation."[22]

CIO, CTO, and e-Government Positions at OMB

Under President George W. Bush, the deputy director of management for OMB served as the federal CIO, providing oversight of agency-level CIOs and coordinating e-government initiatives. In 2001, OMB established the position of associate director for information technology and e-government at OMB to serve as "the leading federal e-government executive," and was given responsibility for the e-government fund, directing the activities of the CIO Council, and advising on the appointments of agency CIOs.[23] The associate director for information technology reported to the OMB deputy director for management. In 2002, Norman Lorentz became the first chief technology officer at OMB, reporting to the associate director for information technology and e-government. In this capacity, Mr. Lorentz was tasked to lead and coordinate multiple efforts to identify and develop the technological architecture needed to support federal e-government and other information technology initiatives.[24]

On April 18, 2009, President Obama announced his appointment of Jeffrey D. Zients to serve as federal chief performance officer (CPO) and OMB deputy director for management.[25] The President previously announced his appointment of Vivek Kundra to serve as federal chief information officer and OMB administrator for e-government and information technology.[26] In announcing the selection of Mr. Chopra, President Obama stated that the CTO would work closely with the federal CPO and CIO "to give all Americans a voice in their government and ensure that they know exactly how we're spending their money – and can hold us accountable for the results."[27]

National Performance Review/National Partnership for Reinventing Government

The Clinton Administration's National Performance Review (NPR)/National Partnership for Reinventing Government (also known as the reinventing government initiative or ReGo) had certain functions similar to ones outlined for a CTO in President Obama's campaign document. In March 1993, President Bill Clinton announced the establishment of the NPR under Vice President Al Gore "to make the entire Federal Government both less expensive and more efficient ... [and to search for] ways to improve services to our citizens and to make our Government work better ... [in part through] better uses of technology."[28] Information technology was one of the tools used to achieve these ends. In 1998, the NPR was renamed the "National Partnership for Reinventing Government" reflecting the administration's intent to shift from "review" to "reinvention," with an emphasis on the use of information technology to transform government operations.[29] In January 2001, the E-Gov website reported

> The National Partnership for Reinventing [Government] urged agencies to use information technology and the Internet to transform how citizens interact with government. Reinvention evolved into e-gov and the goal was to provide better access to government services. Today e-gov is putting people "online, not in line." By the end of 2000, nearly 40 million Americans were doing business with the government electronically. On a regular basis, people are accessing information to solve problems themselves through the Internet, via telephones, and through neighborhood kiosks.[30]

Organizationally, the director of the National Partnership for Reinventing Government served as senior policy advisor to Vice President Gore.[31]

Among ReGo's technology-focused efforts were Access America, which sought the "integration of services across different federal agencies so citizens [could] 'custom-tailor' government to their specific needs," and G-Gov, the use of geographic information and information technology to improve government services.[32]

Department of Commerce Technology Administration

The role of chief government advocate for U.S. civilian technological innovation was previously performed largely by the Commerce Department's Technology Administration (TA) and its predecessors, the Office of Productivity, Technology and Innovation (OPTI)[33] and the Office of Industrial Technology (OIT).[34] During its existence, the Technology Administration was headed by an Under Secretary for Technology and included an analytical arm, the Office of Technology Policy (OTP).[35] In 2007, the Technology Administration, including the positions of Under Secretary for Technology and Assistant Secretary for Technology Policy, were eliminated under the provisions of the America COMPETES Act (P.L. 110-69). The act eliminated the duties of the Technology Administration and did not reassign any of them to other agencies. Among TA's duties:

- to conduct technology policy analyses to improve United States industrial productivity, technology, and innovation;
- to determine the relationships of technological developments and international technology transfers to the output, employment, productivity, and world trade performance;
- to determine the influence of economic, labor and other conditions, industrial structure and management, and government policies on technological developments in particular industrial sectors worldwide;
- to identify technological needs, problems, and opportunities within and across industrial sectors that, if addressed, could make a significant contribution to the economy of the United States;
- to assess whether the capital, technical and other resources being allocated to domestic industrial sectors which are likely to generate new technologies are adequate;

- to propose and support studies and policy experiments to determine the effectiveness of measures with the potential of advancing United States technological innovation;
- to encourage and assist the creation of centers and other joint initiatives by State or local governments, regional organizations, private businesses, institutions of higher education, nonprofit organizations, or Federal laboratories to encourage technology transfer, to stimulate innovation, and to promote an appropriate climate for investment in technology-related industries;
- to propose and encourage cooperative research to promote the common use of resources, to improve training programs and curricula, to stimulate interest in high technology careers, and to encourage the effective dissemination of technology skills within the wider community;
- to serve as a focal point for discussions among United States companies on topics of interest to industry and labor, including discussions regarding manufacturing and discussions regarding emerging technologies; and
- to consider government measures with the potential of advancing United States technological innovation and exploiting innovations of foreign origin.[36]

Some or all of these duties might be assigned to a CTO in the context of serving as chief advocate for technological innovation.

White House Office of Science and Technology Policy

The Office of Science and Technology Policy is the White House office with primary responsibility for science and technology issues.[37] The OSTP traces its organizational roots to the Office of Scientific Research and Development that was established within the Executive Office of the President in 1941 by President Franklin Roosevelt under Executive Order 8807.[38] In 1976, Congress directed the establishment of OSTP in the National Science and Technology Policy, Organization, and Priorities Act (P.L. 94-282), charging the office with serving "as a source of scientific and technological analysis and judgment for the President with respect to major policies, plans, and programs of the Federal Government." The act authorizes OSTP to

- advise the President and others within the Executive Office of the President on the impacts of science and technology on domestic and international affairs;
- lead an interagency effort to develop and implement sound science and technology policies and budgets;
- work with the private sector to ensure Federal investments in science and technology contribute to economic prosperity, environmental quality, and national security;
- build strong partnerships among Federal, State, and local governments, other countries, and the scientific community; and
- evaluate the scale, quality, and effectiveness of the Federal effort in science and technology.

To many observers, President Obama has imbued, in effect, the CTO position with many of the duties of OSTP by selecting a single individual, Mr. Chopra, to serve as both CTO and associate director of OSTP for technology.

POTENTIAL CHALLENGES

Among the early challenges the CTO may face are defining and communicating the roles of the position; identifying and recruiting talent, from both inside and outside of government; and negotiating domains of responsibilities, formal and informal, within the White House and with executive branch agencies that have overlapping missions. Beyond these initial challenges, the CTO will need to establish goals and milestones, set priorities, secure resources, and develop and execute a strategy. If the position or office of the CTO is not established by Congress and provided with statutory authorities and a dedicated budget, it may be difficult for the CTO to affect change in individual federal agencies or systemically throughout the federal government. In such a case, the efficacy of the CTO may depend largely on the mandate provided by President Obama to the CTO (and agencies' perception of the mandate), the imprimatur of the White House, and the personal attributes of the CTO (e.g., relationship with the President, past accomplishments, knowledge, professional reputation, persuasiveness).

Perhaps one of the most difficult and enduring challenges the CTO may face could be "turf wars" associated with overlapping responsibilities with other executive agencies and their principals. These "turf wars" could involve issues such as technology and innovation policy, computer and network security, and intellectual property enforcement. For example, there are several organizations within the Executive Office of the President (EOP) that have potentially overlapping authorities, including

- The Office of Management and Budget (as discussed on pages 6-7 of this report);
- The National Economic Council (NEC), on issues related to technological innovation. For example, the NEC might evaluate federal policies intended to promote technological innovation as an option for achieving the President's economic policy objectives;
- The President's Council on Innovation and Competitiveness, which is charged under the America COMPETES Act with providing advice to the President with respect to global trends in competitiveness and innovation and allocation of Federal resources in education, job training, and technology research and development in the context of global trends in competitiveness and innovation, as well as making recommendations to the heads of executive agencies to improve innovation;[39] and
- The Intellectual Property Enforcement Coordinator, on technology-centered intellectual property rights (e.g., digital copyright).[40]

Potential conflict between the CTO and these offices might be magnified with the co-location of the CTO within OSTP. While assigning Mr. Chopra the dual roles of assistant to the President and CTO and associate director of OSTP may contribute to better coordination and integration of CTO and OSTP functions, it may also create reporting and issue jurisdiction issues, both within OSTP and with other EOP offices. For example, in his capacity as OSTP associate director for technology, Mr. Chopra reports to the director of OSTP, just as the position has in previous administrations. However, in his capacity as assistant to the President and as CTO, both he and the director of OSTP carry the equivalent title of "assistant to the president." In addition, placing the CTO in OSTP may also create jurisdiction issues with the OMB's CPO and CIO functions. The CTO's responsibilities might also overlap with other executive branch agencies, such as the Federal Communications Commission (FCC) and/or the Department of Commerce's National Telecommunications and Information Administration (NTIA) on issues such as information and communication technology, and with agency CIOs and CTOs.

Finally, the manner in which the CTO has been established may affect the position/office's ability to transcend presidential administrations. If the authorities of the CTO continue to rely solely upon the President's executive authority (through an executive order, for example), then its continued existence would be at the sole discretion of the current or future Presidents. In contrast, if the CTO were to be established by statute (as OSTP was, for example), then the position/office would continue to exist through changes of presidential administrations unless eliminated by statute.

CTO Structure and Activities in the Obama Administration

There is no official position description for the CTO,[41] nor has President Obama sought legislation to create a statutory foundation for the CTO position. Accordingly, to date, the structure and official duties of the CTO remain largely undefined.

Structurally, Mr. Chopra has a small staff to assist him in his role as CTO, including three deputy CTOs and an assistant deputy CTO.[42] Appointment to the CTO position is at the sole discretion of the President and does not require Senate confirmation. However, Mr. Chopra's appointment as associate director of OSTP position required Senate confirmation; he was confirmed on May 21, 2009.[43] According to OSTP, the CTO is organizationally a part of OSTP. Further, OSTP states that the CTO and associate director of OSTP for technology are separate and distinct positions though they are currently held by the same individual.[44]

In the absence of a formal position description, some insights into the duties of the CTO in the Obama Administration may be gleaned from what Mr. Chopra has said and done since assuming his dual responsibilities. However, in the course of his public communications, Mr. Chopra does not generally associate his various activities with the unique responsibilities of the two positions he holds. This makes it difficult to identify which of his efforts and comments are specifically linked to his duties as CTO as distinct from those that are linked to his duties as associate director of OSTP.

In a December 2009 interview with a trade publication, Mr. Chopra described his role as CTO as an umbrella over his two-part service as assistant to the President and OSTP associate director for technology:

> The Chief Technology Officer role has two components: the first is my service as Assistant to the President; the second is in my capacity as Associate Director for Technology in the White House Office of Science and Technology Policy. In my role as Assistant to the President, I serve as one of 25 key advisors who meet every morning in the West Wing to

coordinate and collaborate on all the major issues in front of the President. My responsibility is largely to ensure that the voice of technology and innovation is heard across a wide range of policy priorities–healthcare, energy, or education. At OSTP, a role for which the United States Senate confirmed me back in May, I coordinate agency technology R&D and policy and interface with the private sector in advancing our policy priorities.[45]

Mr. Chopra's role as assistant to the President, as articulated above, seems consistent with then-candidate Obama's assertion at a campaign event in Silicon Valley that one of the reasons behind his plan to appoint a CTO is to "make certain we incorporate technology into every decision we make."[46] It is unclear, however, how his self-described role as assistant to the President to "ensure that the voice of technology and innovation is heard across a wide range of policy priorities" aligns with the role of OSTP director John Holdren who also serves as assistant to the President for science and technology. In previous presidential administrations, the role Mr. Chopra asserts for himself has been one generally reserved to the President's science advisor.

According to OSTP,[47] one of Mr. Chopra's responsibilities is to assist in implementation of the President's Strategy for American Innovation, articulated in an Executive Office of the President report published in September 2009.[48] The strategy includes four thrusts:

- "Invest in the Building Blocks of American Innovation," including fundamental research, 21st century workforce skill development, physical infrastructure, and widespread access to information technology tools.
- "Promote Competitive Markets that Spur Productive Entrepreneurship," including promoting American exports, supporting open capital markets, encouraging high-growth and innovation-based entrepreneurship, and improving innovation.
- "Drive Innovations in Health Care Technology," including expanding use of health care information technology, increasing investment in medical research, and slowing the rising costs of health care costs.
- "Catalyze Breakthroughs for National Priorities," including investments in new energy technologies, advanced vehicle technologies, and health care technology, as well as addressing "grand challenges" through science, technology and innovation.[49]

In testimony before the Senate Budget Committee Task Force on Government Performance, Mr. Chopra stated that his focus was on "harnessing the power and potential of technology and innovation to execute on the President's vision for a 21st Century economy" to create jobs, improve U.S. competitiveness, improve the affordability of communications, expand adoption of broadband technologies, help keep families more connected, and increase the safety and security of Americans.[50]

Further, in his testimony, Mr. Chopra discussed three areas of his work:

- Technology for open government. Mr. Chopra cited his efforts in support of the Open Government Directive,[51] issued by the White House Office of Management and Budget, that instructs agencies to provide information to the public in open, accessible, machine-readable formats; requires agencies to formulate an Open Government Plan and website; and calls for a review of government-wide information policies. Mr. Chopra described his office's role in the development of this directive as undertaking a month-long pilot initiative using emerging technologies like wikis, blogs, and posts to solicit ideas from the public on the topic, then culling more than 1,000 responses to incorporate their ideas into the crafting of the directive.

- Technology for government performance. Mr. Chopra discussed several examples of efforts to improve government performance in moving research into development and deployment, supporting open standards, and using prizes and competitions to align innovation with national priorities. It was not clear from his remarks the role the CTO played in these activities.
- The President's Strategy for American Innovation (see earlier discussion).

Mr. Chopra has also advocated for the application of technology and innovation to improve education. In testimony before the House Committee on Education and Labor,[52] Mr. Chopra asserted the need for a greater proportion of Americans to earn college degrees; an increase in the number of students excelling in science, technology, engineering, and mathematics; breakthrough strategies to identify hidden talent in the United States; and the potential role for technology and innovation to help meet theses needs.

It would appear from the initiatives Mr. Chopra has undertaken and the testimony and interviews he has given that his scope of duties is wide and generally aligned with the vision articulated by President Obama during the campaign and transition. Among the tools Mr. Chopra has used to fulfill his CTO responsibilities are the development and promotion of ideas, policy development, public communication, and fostering agency collaboration and coordination.

Issues for Consideration by Congress

With the appointment of Mr. Chopra as CTO, Congress has the opportunity to consider and oversee the Administration's plans for and use of this position. Although the President has provided a broad overview of the role of the CTO, the White House has not provided detailed information on the roles and responsibilities of the CTO, its organizational structure, staffing plans, or relationship to White House offices and federal agencies.[53] Nor has the President sought to create a statutory foundation for the CTO that might be used to define the office's duties and authorities.

On April 2, 2009, Representative Gerald E. Connolly introduced the Chief Technology Officer Act of 2009 (H.R. 1910). Representative Connolly stated that the purpose of the bill, in part, is to make the position of CTO permanent to "ensure that a presidentially-appointed Chief Technology Officer would be part of future administrations."[54] In an interview with the National Journal's nextgov, Representative Connolly further explained,

> What I'm trying to do here is essentially provide a statutory framework for what President Obama has done by executive order.... It is a logical extension of what he has done. It guarantees that the CTO is the spokesman for technology in the federal government and the White House itself—and the advocate for it.[55]

The act would provide a statutory foundation for the Office of the Chief Technology Officer, establishing it as an Executive Office of the President headed by a federal chief technology officer. The act assigns thirteen duties to the office, focused primarily on the Federal government's use of information technology and its role in securing the national information infrastructure. Under the act, the CTO would also "serve as a source of analysis and advice for the President and heads of Federal departments and agencies with respect to major policies, plans, and programs of the Federal Government associated with information technology."[56]

Bruce McConnell, a former head of information technology policy at OMB, asserts that the bill does not give the CTO a broad enough portfolio, placing too great a focus on information technology which instead should be the domain of the federal chief information officer (CIO). Alan Paller of the SANS Institute, a computer security research and education organization, has expressed concerns that the wide range of duties and responsibilities assigned the CTO in the bill might pull the CTO in too many directions and make the CTO ineffectual. Trey Hodgkins, vice president for national security and procurement policy at TechAmerica, an association that represents high-tech companies, stated that it was important for the CTO and CIO to have the authorities to facilitate participatory democracy and, if a statute was necessary to provide these authorities, the legislation may be beneficial.[57]

The act was referred to the House Committee on Oversight and Government Reform on May 4, 2009. No action has been taken.

On May 27, 2010, Representative Ike Skelton offered an amendment (H.Amdt. 658) to the National Defense Authorization Act for Fiscal Year 2011 (H.R. 5136), which included, among other things, Subtitle B, "Federal Chief Technology Officer." The amendment was agreed to in the House by a voice vote. The provisions of Subtitle B are very similar to those in H.R. 1910. The subtitle would establish the Office of the Chief Technology Officer in the Executive Office of the President, and assign to the position a variety of duties related to the federal government's use of information technology (IT). The duties of the CTO would include supporting efforts to foster the further development of IT through research and development funding, development of public-private partnerships to foster federal knowledge and integration of commercially available IT, and collection and analysis of information about IT trends that may affect the federal government's use of IT to meet its goals. The CTO would also conduct analysis and provide advice to the President and heads of federal departments and agencies with respect to major IT policies, plans, and programs of the federal government. Further, the subtitle designates the CTO as a member of the Domestic Policy Council and the Federal Cyber Security Practice Board. It also requires the CTO to produce an annual report for the President and selected committees of Congress that provides information on programs to promote the development of technological innovations; recommendations for the adoption of policies to encourage the generation of technological innovations; and information on the activities and accomplishments of the Office of the Chief Technology Officer.

If Congress chooses to establish the CTO position through statute, there are several issues it may wish to consider. For example:

- What mission, duties, and authorities should be given the CTO? Should one person serve as both the CTO and OSTP associate director for technology?
- What level of funding should be authorized and/or appropriated for the CTO?

- Should the CTO be placed in the Executive Office of the President or elsewhere in the executive branch? If in the EOP, should the CTO directly report to the President, or instead be a part of another EOP agency? Who should the CTO report to? Should the appointment of the CTO be subject to Senate confirmation?
- Should the CTO be a stand-alone position or an office or agency with its own staff? If the CTO is established together with a supporting office or agency, how many full-time equivalent positions should be authorized? What should be the composition of the staff with respect to occupation (e.g., scientists, engineers, IT professionals, lawyers, managers, administrators); political appointees and career civil servants; and permanent employees and employees on detail from other agencies?
- How should the work of the CTO differ, overlap, and/or complement the duties and authorities of offices in the Executive Office of the President, and other executive branch agencies?
- What should be the relationship between the President's CTO and the existing CTOs and CIOs of individual departments and agencies?

End Notes

[1] *Remarks of President Barack Obama*, Weekly Address, April 18, 2009, The White House, available at http://www.whitehouse.gov/the_press_office/Weekly-Address-President-Obama-Discusses-Efforts-to-Reform-Spending-Government-Waste-Names-Chief-Performance-Officer-and-Chief-Technology-Officer/

[2] Personal communication with the White House Office of Science and Technology Policy, April 20, 2009.

[3] "Federal Register," *Executive Order 13500, February 5, 2009*, vol. 74, no. 27 (February 11, 2009), p. 6981.

[4] *Remarks of President Barack Obama*, Weekly Address, April 18, 2009, The White House, available at http://www.whitehouse.gov/the_press_office/Weekly-Address-President-Obama-Discusses-Efforts-to-Reform-Spending-Government-Waste-Names-Chief-Performance-Officer-and-Chief-Technology-Officer/.

[5] Obama '08, *Barack Obama: Connecting and Empowering All Americans Through Technology and Innovation*, November 2007. Available at http://www.barackobama.com/pdf/issues/technology/Fact_Sheet_Innovation_and_Technology.pdf.

[6] Website of the Office of the President-elect. Available at http://change.gov/agenda/technology_agenda/.

[7] Tom Lowry, "The Short List for U.S. Chief Technology Office," *BusinessWeek*, October 19, 2008.

[8] Ibid.

[9] Ibid.

[10] *Change for America: A Progressive Blueprint for the 44th President*, Center for American Progress and New Democracy Project, November 12, 2008.

[11] *Policy Recommendations for President-elect Obama*, Association of American Universities, December 2008, available at http://www.aau.edu/policy/policy_recommendations_new_admin.aspx?id=7710.

[12] Ibid.

[13] The White House, "President Obama Discusses Efforts to Reform Spending, Government Waste; Names Chief Performance Officer and Chief Technology Officer," press release, April 18, 2009, http:// www.whitehouse.gov/ the_press_office/Weekly-Address-President-Obama-Discusses-Efforts-to-Reform-Spending-Government-Waste-Names-Chief-Performance-Officer-and-Chief-Technology-Officer/.

[14] Ibid; also Security and Exchange Commission annual filing, Form 10-K, Advisory Board Company, May 30, 2008, available at http://ccbn.10kwizard.com/xml/download.php?repo=tenk&ipage=5705904&format=PDF.

[15] Chief Information Officers Council website, http://www.cio.gov.

[16] Incorporated as sections D and E of the National Defense Authorization Act for Fiscal Year 1996 (P.L. 104-106)

[17] For additional information, see CRS Report RL34492, *Reauthorization of the E-Government Act: A Brief Overview*, by Jeffrey W. Seifert. This report also provides perspectives on the anticipated benefits and drawbacks of establishing a chief federal CIO position.

[18] Executive Order 13011, "Federal Information Technology," 61 *Federal Register* 37657, July 19, 1996.

[19] P.L. 107-347.

[20] Roger D. Smith, "The Chief Technology Officer: Strategic Responsibilities and Relationships," *Research Technology Management*, July/August 2003.

[21] The close linkage between the duties of a CTO and CIO is illustrated by dual assignments held in some agencies. For example, the Department of Education, National Aeronautics and Space Administration (NASA), and the U.S. Fish and Wildlife Service have CTOs who concurrently hold the title of Deputy CIO. To further illustrate the linkage, at NASA's Goddard Space Flight Center, the CTO is located in the Office of the Chief Information Officer.

[22] Transportation Security Administration, U.S. Department of Transportation, "TSA Names Mike Golden Chief Technology Officer," press release, December 11, 2006, http://www.tsa.gov/press/releases/2006/press_release_12112006.shtm.

[23] Office of Management and Budget, "Mark Forman Named Associate Director for Information Technology and E-Government," press release, 14 June 2001, http://www.whitehouse.gov/omb/pubpress/2001-13.html.

[24] For additional information, see CRS Report RL30914, *Federal Chief Information Officer (CIO): Opportunities and Challenges*, by Jeffrey W. Seifert.

[25] Personal communication with the Office of Management and Budget communications office, April 21, 2009.

[26] The White House, "President Obama Names Vivek Kundra Chief Information Officer," press release, March 5, 2009, http://www.whitehouse.gov/the_press_ office/ President- Obama-Names- Vivek-Kundra-Chief-Information - Officer/; personal communication with the Office of Management and Budget communications office, April 21, 2009.

[27] The White House, "President Obama Discusses Efforts to Reform Spending, Government Waste; Names Chief Performance Officer and Chief Technology Officer," press release, April 18, 2009, http://www. whitehouse. gov/ the_press_office/ Weekly-Address-President-Obama-Discusses-Efforts-to-Reform-Spending-Government-Waste-Names-Chief-Performance-Officer-and-Chief-Technology-Officer/.

[28] *Weekly Compilation of Presidential Documents*, Government Printing Office, vol. 29, Mar. 8, 1993, p. 350. Available at http://frwebgate4.access.gpo.gov/cgi-bin/TEXTgate.cgi? WAISdocID=854991445865+0+1+0 &WAISaction=retrieve.

[29] Remarks of Morley Winograd, director, National Partnership for Reinventing Government, at the Virtual Government '99 Conference, February 23, 1999, available at http://govinfo.library.unt.edu/npr/library/speeches/ 022399.html

[30] The National Partnership for Reinventing Government website, January 2001, as archived at http://govinfo.library.unt.edu/npr/initiati/index.html.

[31] For additional information on the NPR/ National Partnership for Reinventing Government, see CRS Report RL30596, *The National Performance Review and Other Government Reform Initiatives: An Overview, 1993-2001*, by Harold C. Relyea, Maricele J. Cornejo Riemann, and Henry B. Hogue.

[32] Ibid.

[33] OPTI was established by the Federal Technology Transfer Act of 1986 (P.L. 99-502).

[34] OIT was established by the Stevenson-Wydler Technology Innovation Act of 1980 (P.L. 96-480).

[35] TA and OTP were established by the National Institute of Standards and Technology Authorization Act for Fiscal Year 1989 (P.L. 100-519).

[36] These and other duties were assigned to the Technology Administration and its predecessor organizations by the Stevenson-Wydler Technology Innovation Act of 1980 (P.L. 96-480), the Federal Technology Transfer Act of 1986 (P.L. 99-502), and the National Institute of Standards and Technology Authorization Act for Fiscal Year 1989 (P.L. 100-519).

[37] Other White House offices—such as the Council on Environmental Quality, the National Economic Council (NEC), Office of Management and Budget (OMB)—also play important roles in the development, coordination, and implementation of science and technology policies. For additional information about OSTP, see CRS Report RL34736, *The President's Office of Science and Technology Policy (OSTP): Issues for Congress*, by Deborah D. Stine.

[38] Available at http://www.presidency.ucsb.edu/ws/index.php?pid=16137.

[39] The America COMPETES Act (P.L. 110-69) directs the President to establish a President's Council on Innovation and Competitiveness.

[40] The Prioritizing Resources and Organization for Intellectual Property Act of 2008 (P.L. 110-403) directs the President to appoint an Intellectual Property Enforcement Coordinator to serve within the Executive Office of the President.

[41] Telephone conversation with Rick Weiss, assistant director for strategic communications, OSTP, The White House, January 6, 2009.

[42] Mr. Chopra has additional staff support for his role as assistant director of OSTP for Technology. See http://www.ostp.gov/cs/about_ostp/leadership_staff.

[43] See http://www.senate.gov/pagelayout/legislative/one_item_and_teasers/nom_confc.htm.

[44] Ibid.

[45] J.D. Kathuria, "14 Questions for the Federal CTO Aneesh Chopra," *ExecutiveBiz*, December 24, 2009, pp. http://blog.executivebiz.com/exclusive-14-questions-for-federal-cto-aneesh-chopra/6579/comment-page-1.

[46] "Barack Obama expected to be the first US 'Tech President,'" *Agence France-Presse*, November 5, 2008.

[47] Telephone conversation with Rick Weiss, assistant director for strategic communications, OSTP, The White House, January 6, 2009.

[48] Office of Science and Technology Policy/National Economic Council, The White House, *A Strategy for American Innovation: Driving Towards Sustainable Growth and Quality Jobs*, Washington, DC, September 21, 2009, http://www.whitehouse.gov/assets/documents/SEPT_20__Innovation_Whitepaper_FINAL.pdf.

[49] Ibid.

[50] "Data-Driven Performance: Using Technology to Deliver Results," testimony of Aneesh Chopra, Chief Technology Officer and Associate Director for Technology, OSTP, The White House, before the Committee on the Budget, Task Force on Government Performance, U.S. Senate, December 10, 2009.

[51] Open Government Directive, M10-06, Office of Management and Budget, December 8, 2009, http://www.whitehouse.gov/open/documents/open-government-directive. For more information on the Open Government Initiative, see http://www.whitehouse.gov/Open.

[52] "The Future of Learning: How Technology is Transforming Public Schools," testimony of Aneesh Chopra, Chief Technology Officer and Associate Director for Technology, OSTP, The White House, before the Committee on Education and Labor, U.S. House of Representatives, June 16, 2009.

[53] According to the White House Office of Presidential Personnel, the only publicly available information about the CTO (including the roles, responsibilities, structure, organizational placement, and relationship to other offices) is what is available on the White House website. Personal communication between CRS and the White House Office of Presidential Personnel, April 20, 2009.

[54] Rep. Gerald E. Connolly, Extension of Remarks, *Congressional Record*, April 21, 2009, p. E911, http://frwebgate.access.gpo.gov/cgi-bin/getpage.cgi?dbname=2009_record&page=E911&position=all.

[55] Aliya Sternstein, "Congressman pushes to expand chief technology officer's influence," *National Journal's nextgov*, April 22, 2009.

[56] H.R. 1910.

[57] Ibid.

In: Technology Policies and Issues for the U.S. Government ISBN: 978-1-61209-501-1
Editors: Danielle A. Godden and Mason C. Andrews

Chapter 2

U.S. INITIATIVES TO PROMOTE GLOBAL INTERNET FREEDOM: ISSUES, POLICY AND TECHNOLOGY *

Patricial Moloney Figliola, Kennon H. Nakamura, Casey L. Addis and Thomas Lum

SUMMARY

Modern means of communications, led by the Internet, provide a relatively inexpensive, open, easy-entry means of sharing ideas, information, pictures, and text around the world. In a political and human rights context, in closed societies when the more established, formal news media is denied access to or does not report on specified news events, the Internet has become an alternative source of media, and sometimes a means to organize politically.

The openness and the freedom of expression allowed through blogs, social networks, video sharing sites, and other tools of today's communications technology has proven to be an unprecedented and often disruptive force in some closed societies. Governments that seek to maintain their authority and control the ideas and information their citizens receive are often caught in a dilemma: they feel that they need access to the Internet to participate in commerce in the global market and for economic growth and technological development, but fear that allowing open access to the Internet potentially weakens their control over their citizens. The ongoing situation of Google in China is representative of these issues.

Legislation now under consideration in the 111th Congress would mandate that U.S. companies selling Internet technologies and services to repressive countries take actions to combat censorship and protect personally identifiable information. Some believe, however, that technology can offer a complementary and, in some cases, better and more easily implemented solution to some of those issues. They argue that hardware and Internet services, in and of themselves, are neutral elements of the Internet; it is how they are implemented by

* This is an edited, reformatted and augmented version of Congressional Research Service publication, Report R41120, dated September 8, 2010.

various countries that is repressive. Also, Internet services are often tailored for deployment to specific countries; however, such tailoring is done to bring the company in line with the laws of that country, not with the intention of allowing the country to repress and censor its citizenry. In many cases, that tailoring would not raise many questions about free speech and political repression.

This report provides information regarding the role of U.S. and other foreign companies in facilitating Internet censorship by repressive regimes overseas. The report is divided into several sections:

- Examination of repressive policies in China and Iran,
- Relevant U.S. laws,
- U.S. policies to promote Internet freedom,
- Private sector initiatives, and
- Congressional action.

Two appendixes describe technologies and mechanisms for censorship and circumvention of government restrictions.

INTRODUCTION

In the late 1960s and 1970s, advancements in telecommunications technologies enabled the creation of a large-scale, interconnected network called ARPANET ("Advanced Research Projects Agency Network"). ARPANET was created by the Defense Advanced Research Projects Agency as a government-funded enterprise until the mid-1990s, when it began commercialization. Today's Internet is a direct outgrowth of the technologies developed and lessons learned from ARPANET. During the late 1990s, the Internet began having a significant impact on culture and commerce, including the exponential increase of near instant communication by electronic mail (e-mail), text-based discussion forums, and the graphical World Wide Web.

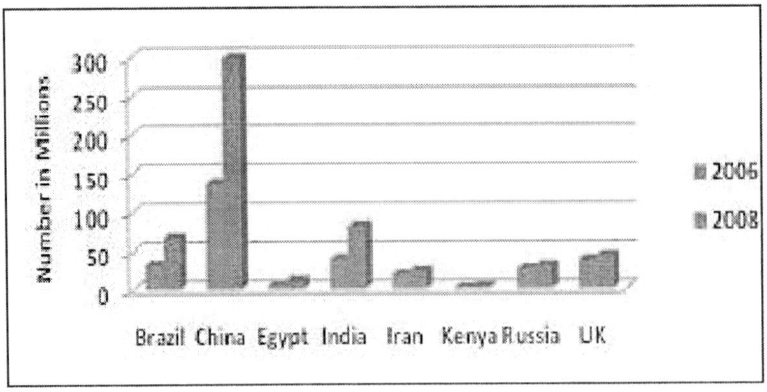

Source: : "Freedom on the Net: A Global Assessment of Internet and Digital Media," Freedom House, April 1, 2009.

Figure 1. Growth in Number of Internet Users in Select Countries 2006-2008

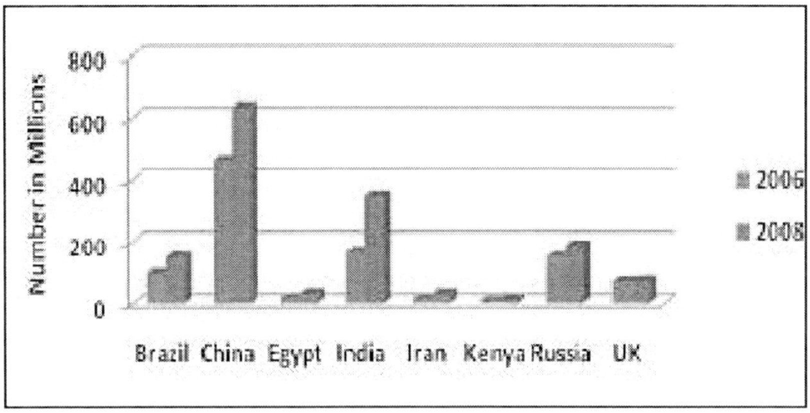

Source: "Freedom on the Net: A Global Assessment of Internet and Digital Media," Freedom House, April 1, 2009.

Figure 2. Growth in Mobile Phone Access in Select Countries 2006-2008

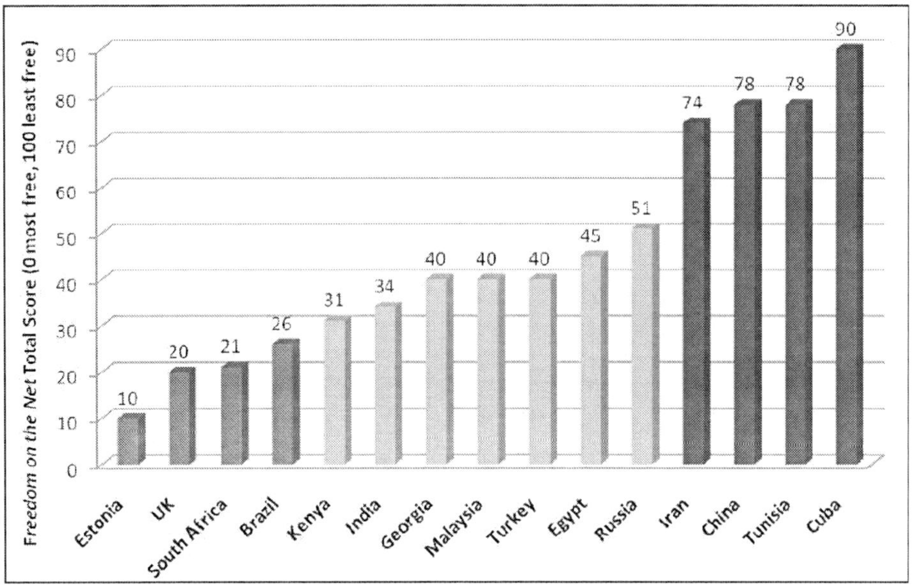

Source: "Freedom on the Net: A Global Assessment of Internet and Digital Media," Freedom House, April 1, 2009

Notes: Estonia to Brazil are "Free." Kenya to Russia are "Partly Free." Iran to Cuba are "Not Free."

Figure 3. Freedom on the Net 15 Country Comparison (0 Best, 100 Worst)

Today, the Internet has evolved even further and many people are using newer tools, such as blogs, social networks, video sharing sites, and other aspects of today's communications technology to express their political ideals, many times in conflict with the political opinions and outlook espoused by their governments. In this way, the Internet has proven to be an unprecedented and often disruptive force in some closed societies, as the governments seek to maintain their authority and control the ideas and information their citizens receive. These regimes are often caught in a dilemma: they need the Internet to participate in commerce in the global market and for economic growth and technological development, but they also seek to restrict the Internet in order to maintain the government's control. **Figure 3** illustrates an assessment by Freedom House[1] of the extent to which selected countries restrict freedom on the Internet.

In Burma during the 2007 Saffron Revolution, YouTube footage, often filmed with cell phone cameras, conveyed to the world the human rights violations against the monks and generated international awareness and reaction. Demonstrations in Tehran following the June 12, 2009, presidential elections were often organized through Twitter and text messages over cell phones.

The Iranian government's violent response to the demonstrations was spread around the world through live cell phone pictures, e-mails, and phone calls. The Voice of America (VOA) reported that during the demonstrations, Iranians sent VOA over 300 videos a day, along with thousands of still pictures, e-mails, and telephone calls to the agency.[2]

A variety of control mechanisms are employed by regimes seeking to limit the ways the Internet is used, ranging from sophisticated surveillance and censorship to threats of retaliation (which foster self-censorship) and actual harassment and arrests of Internet users. Such regimes often require the assistance of foreign Internet companies operating in their countries. These global technology companies find themselves in a dilemma. They often must choose between following the laws and the requests of authorities of the host country, or refusing to do so and risking the loss of business licenses or the ability to sell services in that country. Human rights groups have protested that Yahoo! and Google censor and remove material deemed sensitive by host governments on country-specific search engines.[3] Microsoft is said to censor Chinese versions of its blog platforms.[4] Human rights groups also charge that Yahoo! has provided Chinese authorities personal identifying information about users that has allowed the government to identify and arrest individuals for statements made on the Web.[5] A representative of Google, Inc. acknowledged the problem of government involvement, noting

> As our ... Burma experiences indicate, our products are platforms for free expression, transparency, and accountability. Because of this, we often face efforts by governments throughout the world to restrict or deny access to our products.[6]

The Global Online Freedom Act of 2009 (GOFA) (H.R. 2271), introduced by Representative Christopher Smith, would mandate that companies selling Internet technologies and services to repressive countries take actions to combat censorship and protect personally identifiable information. Some believe, however, that technology can offer a complementary and, in some cases, better and more easily implemented solution to prevent government censorship. Hardware and Internet services, in and of themselves, are neutral elements of the Internet; it is how they are implemented by various countries that makes Internet access "repressive."

For example, hardware, such as routers, is needed to provide Internet service everywhere. However, hardware features intended for day-to-day Internet traffic management, conducted by Internet service providers (ISPs) and governments for benign purposes, can be misused. Repressive governments are able to use these features to censor traffic and monitor use—sometimes using them to identify specific individuals for prosecution. It is not currently feasible to remove those features from the product, even when sold to countries that use those features to repress political speech.[7]

On the other hand, Internet services, such as Google, are often tailored for deployment to specific countries. Such tailoring is done to bring the company's products and services in line with the laws of that country, and not with the end goal of allowing the country to repress and censor its citizenry. In many cases, tailoring does not raise many questions about free speech and political repression because the country is not considered to be a repressive regime. Under Canadian human rights law, for example, it is illegal to promote violence against protected groups; therefore, when reported, Google.ca will remove such links from search results.[8]

Internet censorship and the prosecution of individuals who attempt to circumvent that censorship are unlikely to be eliminated in some countries. However, while some governments are continually looking for new and more thorough methods to restrict or inhibit Internet use, citizens in these countries are active in developing techniques to circumvent those efforts.

EXAMPLES OF COUNTRIES CHARGED WITH RESTRICTING INTERNET FREEDOM

The organization Reporters Without Borders has listed 15 countries where Internet freedom is restricted. These countries are China, Cuba, North Korea, Belarus, Myanmar, Egypt, Ethiopia, Iran, Saudi Arabia, Syria, Tunisia, Turkmenistan, Uzbekistan, Vietnam, and Zimbabwe.[9] This report covers two of these countries, China and Iran, both of which have been in the news during 2009 and 2010.

China[10]

The People's Republic of China (PRC) has the world's largest number of Internet users, estimated at 330 million people, including 70 million bloggers. It also has one of the most sophisticated and aggressive Internet censorship and control regimes in the world. According to some estimates, between 30 and 40 Chinese citizens are serving prison sentences for writing about politically sensitive topics online.[11] In November 2009, Huang Qi, a human rights advocate, was sentenced to three years in prison for "possessing state secrets" after posting online appeals and complaints of families whose children had been killed in school buildings during the Sichuan earthquake of May 2008. Some studies show that the vast majority of Internet users in China do not view the medium as a political tool.[12] Nonetheless, Chinese Internet users are able to access unprecedented amounts of information, despite government attempts to limit the flow, while political activists and others continue to push back against restrictions and find ways to circumvent censorship.

PRC officials have argued that Internet controls are necessary for social stability and that new restrictions target pornography and other "harmful content."[13] Chinese official commentary has suggested that the U.S. government has applied a double standard, regulating the Internet at home while calling for other countries to eliminate controls. The PRC government also has referred to U.S. criticism of Internet restrictions in China as politically motivated and an interference in China's domestic affairs.[14]

The PRC government employs a variety of methods to control online content and expression, including website blocking and keyword filtering; regulating and monitoring Internet service providers, Internet cafes, and university bulletin board systems; registering websites and blogs; and occasional arrests of high-profile "cyber dissidents" or crackdowns on Internet service providers.[15] Some analysts argue that even though the PRC government cannot control all Internet content and use, its selective targeting creates an undercurrent of fear and promotes self-censorship. Blocked websites, social networking sites, and file sharing sites include Radio Free Asia, international human rights websites, many Taiwanese newspapers, Facebook, Twitter, and YouTube. The government reportedly has hired thousands of students to express pro-government views on websites, bulletin boards, and chat rooms.[16] Furthermore, some analysts argue that the Internet has enhanced government propaganda and surveillance capabilities.

Nonetheless, the Internet has made it impossible for the Chinese government to restrict information as fully as before; bulletin boards, comment boards, chat rooms, blogs, and other outlets have allowed for an unprecedented amount of information and public comment on social and other issues. Although the state has the capability to block news of events or to partially shut down the Internet, as it did in Xinjiang following ethnic unrest that erupted there in July 2009, it often cannot do so before such events are publicized, if only fleetingly, online. The threat of public exposure or condemnation through the Internet reportedly has compelled some government officials to conduct affairs more openly. For Chinese Internet users in search of censored information, circumventing government controls is often made possible by way of "proxy servers" or "virtual private networks" using special software.[17] Furthermore, English language news sites, such as the *New York Times* and the *Washington Post*, are generally available.

U.S. Internet Companies, China, and Human Rights Issues

Some human rights activists and U.S. policy makers have expressed concern that U.S. Internet companies have sold Internet services or technologies to China that have assisted the PRC government in restricting information and communication and in monitoring and identifying Internet users. U.S. congressional committees and commissions have held hearings on the topics of global Internet freedom and the roles of U.S. Internet and technology companies in China's censorship regime. Some media watchdog groups and Members of Congress have maintained that some U.S. information technology companies, including Yahoo!, Microsoft, Google, and Cisco Systems, have provided willing, direct, sustained, or comprehensive support to PRC Internet censorship and political control efforts.[18]

U.S. information technology companies have responded that they must abide by the laws of the countries in which they operate, and that they are not actively cooperating or collaborating with the PRC government or tailoring their products to suit PRC censorship requirements.[19] These companies add that despite PRC censorship policies, they nonetheless are enlarging the volume of information available in China and other Internet-restricting countries, and can better press for freedom of expression and protection of privacy while located in these countries. They also claim that Chinese and other Asian and European competitors would fill the void in providing Internet services and technology in their absence. Furthermore, some Chinese experts have suggested that overall, the Internet, including foreign involvement, has created greater political freedom, despite the ongoing battle against growing PRC government attempts to control it.[20]

Yahoo!

Yahoo! has been blamed for complicity in the arrests of at least four Chinese Internet users by providing their e-mail account information to PRC authorities. In the most high-profile case, in 2004, Yahoo!'s Hong Kong office was accused of having provided information about the identity of a Chinese journalist and Yahoo! e-mail account holder, Shi Tao. Shi reportedly had forwarded information about state policy regarding the 15th anniversary of the Tiananmen demonstrations via his Yahoo! e-mail account to an overseas democracy group.[21] In March 2005, a PRC court sentenced Shi to 10 years in prison for "leaking state secrets." In August 2005, Yahoo! bought a 39% stake in China's Alibaba Group, a Chinese Internet service provider, and turned over its PRC operations to the Chinese company.

Microsoft

In 2005, Microsoft shut down the MSN Spaces site of Chinese political blogger Zhao Jing (a.k.a. Michael Anti) at the request of the PRC government, after Zhao had expressed support in his blog for a boycott of *Beijing News* following the firing of one of its editors. Human rights activists also criticized Microsoft for blocking words such as "democracy" from MSN Spaces. Microsoft was China's leading blog service provider at the time and remains one of the most popular. Recently, Microsoft also has been accused of cooperating with China's censorship policies in the development of its new *Bing* search engine.[22]

Google

Google's activities in China have reflected an attempt by the company to comply with PRC policies while limiting the company's role in censorship. Google's Chinese search engine, Google.cn, reportedly is the second-most widely used information-gathering service in China after that of *Baidu*, a Chinese company, and is the least censored, according to one study.[23] Google.cn provides a message stating that a website is unavailable due to "local laws, regulations, and policies," suggesting to the user that additional information exists, but that the government has closed access to that site. In 2006, Google reportedly moved its search records outside of the PRC in order to prevent the government from accessing the data without the company's consent, and does not host Gmail and Blogger services in China as a measure to protect the privacy of Chinese account holders.[24]

Ever since it entered the China market in 2005, Google and the PRC government have clashed over censorship and other issues, although the company has complied with Chinese laws in principle. In June 2009, China's Foreign Ministry accused the Internet company of violating PRC law and enabling Chinese Internet users to access "vulgar content." Google's Chinese service was disrupted for a few days, which some analysts viewed as the Chinese government response to Google's apparent resistance to abide by new censorship edicts.[25] Chinese writers accused Google of copyright infringement after the company began publishing their works in its online library, Google Books.[26] In October 2009, the *People's Daily,* the state's premier newspaper, accused Google of blocking its stories of the dispute.

In February 2010, China accused Google of breaking a "written promise" to follow filter laws regarding searches. In response, Google re-routed searches to uncensored Google Hong Kong—automatically redirecting users from "google.cn" to "google.com.hk." While that allowed Google to be free from China's censorship laws, because it doesn't host search operations on the mainland, Google acknowledged that the Chinese authorities consider the approach "unacceptable." Finally, in June 2010, the day before Google's license to do business in China was due to expire, the company resubmitted a renewal application saying it would create a new "google.cn" landing page, where users could find a link to the non-censored Hong Kong search engine. The tense relationship between China and Google is likely to continue.

Cisco Systems

Cicso Systems, Juniper Networks, Nortel of Canada, and Alcatel of France reportedly were involved in upgrading China's Internet infrastructure, filtering, and surveillance systems earlier this decade. According to some reports, Cisco Systems sold several thousand routers to China, which helped to facilitate the PRC government's censorship of Internet content and monitoring of Internet users.[27] According to other reports, Cisco sold technology to China's police force that can be used in the collection and use of data regarding personal background and imaging information, Web browsing history, and e-mail.[28]

The Continuing Battle between Censorship and Freedom of Information

The PRC government has displayed a growing nervousness about the Internet's influence on Chinese society and politics, but it has been reluctant to provoke the ire of China's online population or to reduce the attractiveness of China's business environment for foreign investors. In June 2009, the PRC government issued a directive requiring "Green Dam Youth Escort" software, designed to prevent children from accessing "harmful content," such as pornography, on all Chinese computers sold after July 1, 2009, including those imported from abroad. Many Chinese Internet users, international human rights activists, foreign governments, chambers of commerce, and information technology manufacturers openly opposed the policy, arguing that the software would undermine computer operability, that it could be used to expand censorship to include political content, and that it could incorporate pirated software and weaken Internet security.[29] On June 30, 2009, the PRC government announced that mandatory installation of the software would be delayed for an indefinite period. On August 14, 2009, Minister of Industry and Information Technology Li Yizhong stated that the directive had created misunderstandings and that, "We will listen to the public's views before issuing a new directive on Green Dam."[30]

Following the aborted launch of "Green Dam," the PRC government has continued to tighten controls over Internet content and use, but in a quieter manner. In September 2009, PRC authorities issued requirements that new users register their true identities. This regulation reportedly has not been well enforced; however, the government can still track down individuals through their IP addresses. In December 2009, new restrictions aimed at cracking down on pornography, media piracy, and threats to national security and stability resulted in the closing of hundreds of websites, many of them entertainment-oriented. Furthermore, the China Internet Network Information Center announced that individuals could no longer apply for ".cn" domain names (China's country code), which it would now limit to registered business enterprises. Some observers argued that these policies could dampen the richness and vibrancy of Internet content and activity in China, as well as provoke a public backlash.[31] On October 15, 2009 (Internet Human Rights Day), 15 Chinese intellectuals issued a Declaration of Internet Human Rights calling for freedom of opinion, speech, and publication online.[32]

Google and Cyber Attacks

In January 2010, Google threatened to cease censoring its Chinese search engine or to pull out of China. The company asserted that, in December 2009, Chinese hackers had attacked its Gmail service and corporate network as well as the computer systems of many other large U.S. corporations in the PRC.[33] Hackers appeared to have targeted the Gmail accounts of Chinese human rights activists; the intellectual property, including "source codes" or programming languages, of Google and other companies; and information on U.S. weapons systems. In a statement, Google's chief legal officer announced that the company would no longer censor results on Google.cn, even if that meant having to shut down the search engine, and potentially its offices in China.[34] Yahoo!, which was also hit by Chinese hackers, expressed support for Google's actions, thereby provoking an angry response by its PRC partner, *Alibaba*.

Chinese discussion boards and micro-blog postings indicated that a small majority of China's online population—and perhaps a large majority of its most active Internet users—wanted Google to stay in China, with some supporting Google's challenge to the PRC government. A significant minority adopted a pro-government stance or interpreted Google's move as profit-oriented.[35] According to some analysts, although China has huge potential, the company currently earns an estimated $300 million to $400 million from its China operations, a "tiny fraction" of its $22 billion in sales worldwide.[36]

While visiting Shanghai during his state visit to China in November 2009, President Barack Obama expressed support of unrestricted Internet access and disapproval of censorship. On January 21, 2010, in a policy speech on Internet freedom, Secretary of State Hillary Clinton urged U.S. Internet companies to oppose censorship in their overseas operations and announced that the Global Internet Freedom Taskforce (GIFT) would be reinvigorated. She also called upon the PRC government to conduct a thorough investigation of the December 2009 cyberattacks upon U.S. companies in China and to make its results transparent. Although Google confirmed in February that the attacks had come from mainland China, Beijing denied involvement in the attacks and defended its Internet policies. The Foreign Ministry stated that foreign companies, including Google, "should respect the laws and regulations, respect the public interest of Chinese people and China's culture and customs and shoulder due social responsibilities."[37]

Iran[38]

The Iranian government has restricted Internet usage since access spread beyond universities and government agencies to the general population in the late 1990s. Today, Iran has an estimated 23 million Internet users,[39] and watchdog groups and Internet activists claim that Iran's filtering and monitoring of usage is among the most extensive in the world.

Additional information about Iran's Internet policies is available from the U.S. Department of State in its annual report, *2008 Country Reports on Human Rights Practices*.

The Iranian government tracks online communication and content through a centralized location in the state's telecommunications monopoly, the Ministry of Communications and Information Technology (MCIT). In addition to its 23 million Internet users, the Persian blogosphere is among the world's most robust. The status of Internet sites and blogs remains contested under Iranian law, but the Press Law does require that bloggers obtain licenses, and all content on websites and blogs is subject to approval of the Ministry of Culture and Islamic Guidance (MCIG). The government also regulates access to the Internet by limiting the speed of Internet access that ISPs can provide to households and public access sites (Internet cafes) to 128 kilobytes per second, making it difficult or impossible to download multimedia content. Iran reportedly is the only country to have imposed a cap on Internet access speed for households.[40] Iran also has arrested numerous activists, bloggers, and journalists on charges of "antigovernment publicity," "propaganda against the Islamic Republic," and "jeopardizing national security."[41]

The government has disabled the Internet altogether in the past, usually during elections, but some observers argue that improvements in monitoring and filtering technologies have made such measures unnecessary and even enabled the government to use the Internet to disseminate disinformation and pro-government content. Following the disputed 2009 presidential election, the Internet was reportedly slow but accessible. The number of detentions of Internet activists and bloggers increased during the post-election unrest, arguably demonstrating the extent of government filtering and monitoring of usage. The post-election crackdown on Internet freedom raised concerns that Iran's human rights abuses were being aided by Western technology companies. Others said the concerns were being overstated, asserting that Iran also develops its own filtering and monitoring technologies.

The Nokia Siemens Network (NSN)[42] sold communication monitoring equipment to the Iranian government in 2008.[43] The monitoring center, installed into the MCIT gateway, was part of a larger contract with Iran that included mobile phone network technology. The Iranian government had reportedly experimented with the monitoring equipment prior to the election, but did not use it extensively until after the election. Some experts have argued that the nature of the content inspection happening in Iran since the election goes beyond the practices of other countries, including China.[44]

NSN maintains that it sold the technology for the purpose of "lawful intercept" of information used to track criminals and terrorists.[45] Critics argue that in a country like Iran, where the population is heavily reliant on Internet communication with the outside world due to censorship of other communication, this technology enables the government to intensify repression.[46]

U.S. Law and Internet Freedom Abroad[47]

In response to laws and regulations of foreign countries requiring censorship and disclosure of users' personal information, some U.S. technology firms engage in Internet censoring and filtering. Some examples include China and other Internet-restricting countries such as Iran. In some cases, such as in Iran, Internet censoring and filtering reportedly involve a practice often called deep packet inspection which is under a great deal of scrutiny in the United States.[48] Doing business in a foreign country subjects the business to the jurisdiction of that country.[49] Nonetheless, concerns have been raised that China's Internet filtering could run afoul of world trade obligations.[50]

U.S. Policy for the Promotion of Internet Freedom Abroad[51]

The importance of Internet freedom to the United States was declared in 2006. During an explanation of that year's State Department 2006 *Country Reports on Human Rights Practices*, then-Under Secretary of State for Democracy and Global Affairs Paula J. Dobriansky explained that the 2006 reports included new, additional focus on "the extent to which internet access is available to and used by citizens in each country and ... whether governments inappropriately limit or block access to the internet or censor websites."[52] This was added as an area of concern because the internet is playing a growing role in people's ability to freely express themselves and in the free flow of information. In discussing this new area of focus, then-Under Secretary Dobriansky said,

> We will continue to defend internet freedom, including by addressing internet repression directly with the foreign governments involved and seeking to persuade foreign officials that restricting internet freedom is contrary to their own interests and that of their countries. The new information in this year's reports will make an important contribution.[53]

At this same time, then-Secretary of State Condoleezza Rice also established the Global Internet Freedom Task Force (GIFT) in order to provide a U.S. foreign policy response to violations of Internet freedom by repressive regimes around the world.[54]

Secretary of State Hillary Rodham Clinton, in a January 21, 2010, speech, stated that Internet freedom is a central part of U.S. foreign policy. She stated that Internet freedom is more than a question of information freedom, it is about the nature of the world we want to inhabit. Clinton further stated: "It's about whether we live on a planet with one Internet, one global community, and a common body of knowledge that benefits and unites us all, or a fragmented planet in which access to information and opportunity is dependent on where you live and the whims of censors."[55]

In her remarks, Secretary Clinton placed the United States on the side of a single Internet where everyone has equal access to knowledge and ideas. She noted that blogs, e-mails, social networks, and text messages are opening up a new virtual town square where citizens can go to criticize their governments and exchange ideas. U.S. responsibility to support this new "town square" is not new but can be found in the First Amendment of the U.S. Bill of Rights ensuring freedom of speech, assembly, and religion. Secretary Clinton argued that these principles were reaffirmed in President Franklin Roosevelt's "The Four Freedoms" speech,[56] and in the work of the United States and its support of the Universal Declaration of Human Rights.

Secretary Clinton further explained that U.S. foreign policy is premised on the idea that no country stands to benefit more than the United States when there is cooperation among peoples and states. No country shoulders a heavier burden than the United States when conflict and misunderstanding make the international system unstable, and force people and countries apart. She stated that it is important that the United States seizes the opportunities that come with interconnectivity and work for a world in which access to networks and information brings people closer together and expands the definition of the global community.

Secretary Clinton continued the GIFT and its responsibilities. The Task Force is co-chaired by the Under Secretaries of State for Democracy and Global Affairs and for Economic, Business, and Agricultural Affairs and draws on the State Department's multidisciplinary expertise in its regional and functional bureaus to work on issues such as international communications, human rights, democratization, business advocacy and corporate social responsibility, and country specific concerns. The task force supports Internet freedom by[57]

- monitoring Internet freedom and reporting in its annual *Country Reports on Human Rights Practices* the quality of Internet freedom in each country around the world;
- responding in both bilateral and international fora to support Internet freedom; and
- expanding access to the Internet with greater technical and financial support for increasing availability of the Internet in the developing world.

In advancing Internet freedom as an objective of U.S. foreign policy, Secretary Clinton proposed a number of key initiatives:[58]

- Continue the work of the State Department's GIFT as it oversees U.S. efforts in more than 40 countries to help individuals circumvent politically motivated censorship by developing new tools and providing the training needed to safely access the Internet;
- Make Internet freedom an issue at the United Nations and the U.N. Human Rights Council in order to enlist world opinion and support for Internet Freedom;
- Work with new partners in industry, academia, and non-governmental organizations to establish a standing effort to advance the power of "connection technologies" that will empower citizens and leverage U.S. traditional diplomacy;
- Provide new, competitive grants for ideas and applications that help break down communications barriers, overcome illiteracy, and connect people to servers and information they need;
- Urge and work with U.S. media companies to take a proactive role in challenging foreign governments' demands for censorship and surveillance; and
- Encourage the voluntary work of the communications-oriented, private sector-led Global Network Initiative (GNI). The GNI brings technology companies, nongovernmental organizations, academic experts, and social investment funds together to develop responses and mechanisms to government requests for censorship.

To fund U.S. efforts in support of Internet freedom, Congress in FY2008 appropriated \$15 million, most of which has been spent or is obligated. Another \$5 million was appropriated in FY2009. Finally, in Secretary Clinton's January 21 speech, she spoke of an additional \$15 million for FY2010 that has been allocated from State Department appropriations to a range of programs that, in full or in part, support Internet freedom.

Assistant Secretary for Democracy, Human Rights, and Labor Michael Posner describes these programs as "not just circumvention.... [I]t's a lot about training people.... It's some about technology. It's some about encouraging groups that are in danger. It's a lot about diplomacy, too, for us getting out there and being sure that when groups are in trouble, we provide a lifeline."[59]

The U.S. Broadcasting Board of Governors' International Broadcasting Bureau also supports counter-censorship technologies and has committed approximately $2 million per year to help enable Internet users in repressive regimes to have access to the VOA and other U.S. governmental and non-governmental websites and to receive VOA e-mail newsletters.

Some observers have expressed concerns that there could be serious negative consequences for U.S. and foreign companies, and U.S. or foreign nationals working or living in countries with repressive regimes, if they follow the expanded U.S. policy supporting Internet freedom. These commenters point out that repressive governments could punish or make an example of an individual or company for not following the dictates of that country. Such actions could include harassment, lifting of business licenses, confiscation of assets, or imprisonment. These observers question what powers, beyond expressing U.S. displeasure through official demarches and public statements or through negotiations, that the United States may have to respond to such actions.[60]

Congressional Action

In 2010, Congress has taken steps to address ongoing concerns about ensuring the free and secure flow of information over the Internet:

- On March 10, 2010, the House Committee on Foreign Affairs conducted a hearing, "The Google Predicament: Transforming U.S. Cyberspace Policy to Advance Democracy, Security, and Trade," on the December 2009 Chinese cyber attacks on Google and other U.S. companies, to consider policy tools to address Internet freedom, trade, and cyber security issues;
- On March 9, 2010, Representatives David Wu and Christopher Smith announced the formation of the House Global Internet Freedom Caucus; and
- The Senate Judiciary Committee, Subcommittee on Human Rights and the Law, held a hearing on March 2, 2010, entitled "Global Internet Freedom and the Rule of Law, Part II" to examine human rights, corporate responsibility, and other issues related to Internet censorship around the world.

THE GLOBAL NETWORK INITIATIVE: PRIVATE SECTOR SUPPORT OF INTERNET FREEDOM[61]

The Global Network Initiative (GNI) was formed in October 2008 to respond to criticism of Internet service providers and computer manufacturers who had sold technology or services to Internet-restricting countries.[62] GNI was launched by a coalition of human rights organizations, academics, investors and technology leaders. GNI adopts a self-regulatory approach to protect and advance individuals' rights to free expression and privacy on the Internet. A set of principles and supporting mechanisms provide guidance to the information and communications technology (ICT) industry and its stakeholders on how to protect and advance the human rights of freedom of expression and privacy when faced with pressures from governments to take actions that infringe upon these rights.

Governments are not members of the GNI, but are encouraged to support the principles and encourage their adoption. Organizations participating in the GNI include Google Inc., Microsoft Corp., and Yahoo! Inc. Each initial participating company committed $100,000 per year over the two-year start-up period. Organizations not participating in the initiative who where involved in its development include Amnesty International and Reporters Without Borders. Reporters Without Borders remains skeptical about how much change GNI can effect, and pushed for standards that would require all government requests and takedown notices be made in writing.

The GNI's Principles on Freedom of Expression and Privacy ("the Principles") are based on internationally recognized laws and standards for human rights, including the Universal Declaration of Human Rights (UDHR), the International Covenant on Civil and Political Rights (ICCPR) and the International Covenant on Economic, Social, and Cultural Rights.

The GNI acknowledges that the rights of privacy and of freedom of expression should not be restricted by governments, except in narrowly defined circumstances based on internationally recognized laws or standards. The Implementation Guidelines ("The Guidelines") of the GNI provide guidance to ICT companies on how to implement the Principles, and describe the actions that constitute compliance.

With respect to government demands to remove or limit access to content or restrict communications, participating companies commit to encourage governments to

- be specific, transparent, and consistent in the demands issued to restrict freedom of expression online;
- encourage government demands that are consistent with international laws and standards;
- require governments to follow local legal processes, interpret government demands so as to minimize the negative effect, when required to restrict communications or remove content; and
- interpret the governmental authority's jurisdiction to minimize the negative effect.

Participating companies commit to operate in a transparent manner when required to remove content or restrict access, and must disclose to users the applicable laws and policies requiring such action, the company's policies for responding to government demands, and provide timely notice to users when access to content has been locked or communications limited due to government restrictions. With respect to privacy, participating companies commit to assess the human rights risks associated with the collection, storage, and retention of personal information and to develop mitigation strategies.

A system of independent third-party assessment of company compliance with the Principles and Implementation Guidelines will be phased in over three stages:

- In Phase One (ends December 2010) each participating company establishes internal policies and procedures to implement the Principles, and the Board approves independence and competence criteria for the selection of independent assessors.

- In Phase Two (2011) independent assessors will conduct process assessments of each participating company to review and evaluate their internal systems for implementing the Principles.
- In Phase Three (January 2012 onwards) the Board will accredit independent assessors to review the internal systems of companies, and company responses to specific government demands implicating freedom of expression or privacy. Each participating company will submit an annual report to the Organization. The assessors will prepare reports explaining each company's responses to government demands, evaluating the effectiveness of the company's responses. Each company will be given the opportunity to respond to the assessor's draft and final report. The Board of the Organization will assess whether the company is in compliance with the Principles and its determination will be made public. The Board of the Organization will publish an annual report assessing each participating company's compliance with the Principles.

RECENT LEGISLATIVE ACTION[63]

Public Laws

H.R. 2647, *National Defense Authorization Act for Fiscal Year 2010*. Introduced by Representative Skelton (by request), referred to the House Armed Services Committee. Enacted October 28, 2009, P.L. 111-84.

Title XII: Matters Relating to Foreign Nations

Subtitle D: VOICE Act—Victims of Iranian Censorship Act or VOICE Act

(Sec. 1242) Expresses the sense of the Senate in support of the universal values of freedom of speech, the press, and expression as it pertains to the people of Iran, and condemns acts of censorship, intimidation, and other restrictions on such freedom in Iran.

(Sec. 1243) States that it shall be the policy of the United States to (1) support freedom of the press, speech, expression, and assembly in Iran; (2) support the Iranian people as they seek, receive, and impart information and promote ideas in writing, print, and through other media; (3) discourage businesses from aiding efforts to interfere with the ability of the Iranian people to access or share information or otherwise infringe upon such freedoms; and (4) encourage the development of technologies that facilitate efforts of the Iranian people to share such information, exercise such freedoms, and engage in Internet-based education programs and other exchanges between U.S. citizens and Iranians.

(Sec. 1244) Authorizes appropriations for the (1) International Broadcasting Operations Fund to expand Farsi language programming and to disseminate accurate and independent information to the Iranian people through radio, television, Internet, cellular telephone, short message service, and other communications; and (2) Broadcasting Capital Improvements Fund to expand transmissions of Farsi language programs to Iran.

(Sec. 1245) Establishes in the Treasury the Iranian Electronic Education, Exchange, and Media Fund to support the development of technologies that will aid the Iranian people in exchanging information and exercising freedom of speech, expression, and assembly. Authorizes appropriations to the Fund.

(Sec. 1246) Directs the President to report annually to Congress on the use of funds authorized under this Subtitle.

(Sec. 1247) Requires the President to (1) direct the appropriate officials to examine claims that non-Iranian companies have provided hardware, software, or other forms of assistance to the government of Iran that has furthered its efforts to filter online political content, disrupt cell phone and Internet communications, and monitor the online activities of Iranian citizens; and (2) report study results to Congress.

(Sec. 1248) Authorizes appropriations to the Secretary of State to document, collect, and disseminate information about human rights in Iran, including abuses since the Iranian presidential election on June 12, 2009.

Bills and Resolutions in the House of Representatives

H.R. 2271, *Global Online Freedom Act of 2009.* Introduced by Representative Christopher Smith and referred to the House Committee on Foreign Affairs; and the House Committee on Energy and Commerce.

Makes it U.S. policy to (1) promote the freedom to seek, receive, and impart information and ideas through any media; (2) use all appropriate instruments of U.S. influence to support the free flow of information without interference or discrimination; and (3) deter U.S. businesses from cooperating with Internet-restricting countries in effecting online censorship.

Expresses the sense of Congress that (1) the President should seek international agreements to protect Internet freedom; and (2) some U.S. businesses, in assisting foreign governments to restrict online access to U.S.-supported websites and government reports and to identify individual Internet users, are working contrary to U.S. foreign policy interests.

Amends the Foreign Assistance Act of 1961 to require assessments of electronic information freedom in each foreign country.

Establishes in the Department of State the Office of Global Internet Freedom (OGIF).

Directs the Secretary of State to annually designate Internet-restricting countries. Prohibits, subject to waiver, U.S. businesses that provide to the public a commercial Internet search engine, communications services, or hosting services from locating, in such countries, any personally identifiable information used to establish or maintain an Internet services account.

Requires U.S. businesses that collect or obtain personally identifiable information through the Internet to notify the OGIF and the Attorney General before responding to a disclosure request from an Internet-restricting country. Authorizes the Attorney General to prohibit a business from complying with the request, except for legitimate foreign law enforcement purposes.

Requires U.S. businesses to report to the OGIF certain Internet censorship information involving Internet-restricting countries.

Prohibits U.S. businesses that maintain Internet content hosting services from jamming U.S.-supported websites or U.S.-supported content in Internet-restricting countries.

Authorizes the President to waive provisions of this act: (1) to further the purposes of this act; (2) if a country ceases restrictive activity; or (3) if it is the national interest of the United States.

H.R. 4784,[64] *Internet Freedom Act of 2010.* Introduced by Representative Wu and referred to the House Science and Technology Committee, Subcommittee on Research and Science Education.

Directs the National Science Foundation to establish the Internet Freedom Foundation governed by a board of 12 members, with equal representation from government, academia, and the private sector. The Internet Freedom Foundation shall—

- Award competitive, merit-reviewed grants, cooperative agreements, or contracts to private industry, universities, and other research and development organizations to develop deployable technologies to defeat Internet suppression and censorship; and
- Award incentive prizes to private industry, universities, and other research and development organizations that successfully develop deployable technologies to defeat Internet suppression and censorship.

The Internet Freedom Foundation shall be funded by such sums as may be necessary.

H.Res. 590, *Expressing grave concerns about the sweeping censorship, privacy, and cybersecurity implications of China's Green Dam filtering software, and urging U.S. high-tech companies to promote the Internet as a tool for transparency, freedom of expression, and citizen empowerment around the world.* Introduced by Representative Wu and referred to the House Committee on Foreign Affairs.

Expresses (1) grave concerns about the sweeping censorship, privacy, and cybersecurity implications of China's Green Dam filtering software; and (2) support for the Chinese people in their quest for Internet freedom and free expression.

Calls on (1) the Chinese government to rescind its requirement for Green Dam to be preinstalled on all new computers; and (2) U.S. high-tech companies to promote the Internet as a tool for transparency, freedom of expression, and citizen empowerment around the world.

H.Res. 672, *Calling on the Government of the Socialist Republic of Vietnam to release imprisoned bloggers and respect Internet freedom.* Introduced by Representative Sanchez and referred to the House Committee on Foreign Affairs. Passed on October 21, 2009.

Supports the right of the citizens of the Socialist Republic of Vietnam to access websites of their choosing and to have the freedom to share and publish information over the Internet.

Calls on Vietnam to (1) repeal Circular 07, Article 88, and similar statutes that restrict the Internet, so as to be in line with the International Covenant on Civil and Political Rights, to which Vietnam is a signatory; (2) become a responsible member state of the international community by respecting individuals' freedom of speech, freedom of press, and freedom of political association; and (3) release all political prisoners, including but not limited to 18 named bloggers and cyber activists.

Appendix A. Technologies Used to Monitor and Censor Web Sites and Web-Based Communications[65]

Key-Word List Blocking

This is a simple type of filtration where a government drops any Internet packets featuring certain keywords, such as "protest" or "proxy."

Domain Name System (DNS) Poisoning

DNS poisoning intentionally introduces errors into the Internet's directory service to misdirect the original request to another IP address.

IP Blocking

IP Blocking is one of the most basic methods that governments use for censorship, as it simply prevents all packets going to or from targeted IP addresses. This is an easy technology to implement, but it does not address the problem of individual communications between users. This method is used to block banned websites, including news sites and proxy servers that would allow access to banned content, from being viewed.

Bandwidth Throttling

Bandwidth throttling simply limits the amount of traffic that can be sent over the Internet. Keeping data volume low facilitates other methods of monitoring and filtering by limiting the amount of data present.

Traffic Classification

This is a much more sophisticated method of blocking traffic than IP blocking, as governments can halt any file sent through a certain type of protocol, such as FTP. Because the government knows that FTP transfers are most often sent through TCP port 21, they can simply limit the bandwidth available on that port and throttle transfers. This type of traffic-shaping practice is the most common one used by repressive governments today. It is not resource intensive and it is fairly easy to implement.

Shallow Packet Inspection (SPI)

Shallow packet inspection is a less sophisticated version of the deep packet inspection (DPI) technique (DPI is described below) that is used to block packets based on their content. Unlike DPI, which intercepts packets and inspects their fingerprints (fingerprinting is described below), headers, and payloads, SPI makes broad generalities about traffic based solely on evaluating the packet header. Although shallow packet inspection can't provide the same refined/detailed traffic assessments as DPI, it is much better at handling volume than DPI.

SPI is much less refined than DPI, but it is capable of handling a greater volume of traffic much more quickly. SPI is akin to judging a book by its cover. This method is prone to exploitation by users because they can disguise their packets to look like a different kind of traffic.

Packet Fingerprinting

This is a slightly more refined method of throttling packets than shallow packet inspection, as it looks not only at the packet header but at its length, frequency of transmission, and other characteristics to make a rough determination of its content. In this manner, the government can better classify packets and not throttle traffic sent out by key businesses.

Deep Packet Inspection (DPI) / Packet Content Filtering

DPI is the most refined method that governments have for blocking Internet traffic. As mentioned above, deep packet inspectors examine not only a packet's header but also its payload. For instance, certain keywords can be both monitored and the e-mail containing them can be kept from reaching its intended destination.

This gives governments the ability to filter packets at a more surgical level than any of the other techniques discussed so far. While providing the most targeted traffic monitoring and shaping capabilities, DPI is also more complicated to run and is far more labor-intensive than other traffic-shaping technologies.

Appendix B. Technologies Used to Circumvent Censorship[66]

Each of the circumvention methods explained below can, in general, be considered an anonymous "proxy server." A proxy server is a computer system or an application program that acts as an intermediary for requests from a user seeking resources from other servers, allowing the user to block access to his or her identity and become anonymous.

Web-Based Circumvention Systems

Web-based circumvention systems are special web pages that allow users to submit a URL and have the web-based circumventor retrieve the requested web page. There is no connection between the user and the requested website as the circumventor transparently proxies the request allowing the user to browse blocked websites seamlessly. Since the web addresses of public circumventors are widely known, most Internet filtering applications already have these services on their block lists, as do many countries that filter at the national level.

Examples: Proxify, StupidCensorhip, CGIProxy, psiphon, Peacefire/Circumventor.

Web and Application Tunneling Software

Tunneling encapsulates one form of traffic inside of other forms of traffic. Typically, insecure, unencrypted traffic is tunneled within an encrypted connection. The normal services on the user's computer are available, but run through the tunnel to the non-filtered computer which forwards the user's requests and their responses transparently. Users with contacts in a non-filtered country can set up private tunneling services while those without contacts can purchase commercial tunneling services. "Web" tunneling software restricts the tunneling to web traffic so that web browsers will function securely, but not other applications. "Application" tunneling software allows the user to tunnel multiple Internet applications, such as e-mail and instant messenger applications.

Examples: Web Tunneling: UltraReach, FreeGate, Anonymizer, Ghost Surf. Examples: Application Tunneling: GPass, HTTP Tunnel, Relakks, Guardster/SSH.

Anonymous Communications Systems

Anonymous technologies conceal a user's IP address from the server hosting the website visited by the user. Some, but not all, anonymous technologies conceal the user's IP address from the anonymizing service itself and encrypt the traffic between the user and the service. Since users of anonymous technologies make requests for web content through a proxy service, instead of to the server hosting the content directly, anonymous technologies can be a useful way to bypass Internet censorship. However, some anonymous technologies require users to download software and can be easily blocked by authorities.

Examples: Tor, JAP ANON, I2P

End Notes

[1] Freedom House is an independent watchdog organization that supports the expansion of freedom around the world. Freedom House supports democratic change, monitors freedom, and advocates for democracy and human rights. More information can be found on its website, http://www.freedomhouse.com.

[2] Danforth Austin, Director, Voice of America, testimony before the Subcommittee on Europe, House Committee on Foreign Affairs, Washington, July 23, 2009.

[3] Lucie Morillon, Washington Director of Reporters Without Borders, Testimony before the Tom Lantos Human Rights Commission, U.S. House of Representatives, Washington, June 18, 2009.
[4] Ibid.
[5] Ibid.
[6] Nicole Wong, Deputy General Counsel, Google, Inc., Testimony before the U.S. Senate Judiciary Committee's Subcommittee on Human Rights and the Law, Washington, May 20, 2008.
[7] Testimony of Mark Chandler, Cisco Systems, before the Senate Committee on the Judiciary Subcommittee on Human Rights and the Law, May 2, 2008.
[8] Testimony of Nicole Wong, Google, op. cit. May 2, 2008.
[9] See Reporters Without Borders, "Handbook for Bloggers and Cyber-Dissidents," http://www.rsf.org/IMG/pdf/guide_gb_md-2.pdf.
[10] Prepared by Thomas Lum, Specialist in Asian Affairs, 7-7616.
[11] U.S. Department of State, 2008 *Human Rights Report: China*, February 25, 2009; PEN American Center, "Failing to Deliver: An Olympic-Year Report Card on Free Expression in China," July 8, 2008.
[12] Rebecca MacKinnon, "Bloggers and Censors: Chinese Media in the Internet Age," *China Studies Center*, May 18, 2007.
[13] Kim Zetter, "China Stands Firm in Response to Google Threat," *Wired*, January 14, 2010.
[14] Gillian Wong, "China Denies Involvement in Google Hackings," *Associated Press*, January 25, 2010.
[15] Some experts estimate that the PRC government has employed 30,000 "Internet police." "On the Wrong Side of Great Firewall of China," *New Zealand Herald*, November 27, 2007.
[16] David Bandurski, "China's Guerrilla War for the Web," *Far Eastern Economic Review*, Vol. 171, no. 5 (July/ August 2008).
[17] Such software is available internally and through foreign sources, including the U.S. government.
[18] The Tom Lantos Human Rights Commission, "The State of Global Internet Freedom," June 18, 2009; U.S. Congress, Senate Committee on the Judiciary, Subcommittee on Human Rights and the Law, *Global Internet Freedom: Corporate Responsibility and the Rule of Law*, May 20, 2008. U.S. Congress, House Committee on International Relations, Subcommittee on Africa, Global Human Rights and International Operations and Subcommittee on Asia and the Pacific, *The Internet in China: A Tool For Freedom or Suppression?*, February 15, 2006.
[19] Cisco's general counsel argued that Cisco does not customize its equipment for China; filtering technologies that are intrinsic to Cisco products cannot feasibly be eliminated; Cisco has a written code of conduct that aims to prevent the modification of its products in foreign countries in such as way as to undermine human rights; and Cisco complies with all U.S. government regulations or export controls that restrict the sale of high tech products and crime detection equipment. See Anne Broache, "Senators Weigh New Laws over China Online Censorship," news.cnet.com, May 20, 2008; Mark Chandler, Cisco Systems, Testimony before the Senate Committee on the Judiciary, Subcommittee on Human Rights and the Law, May 20, 2008; Mark Chandler, Cisco Systems, Testimony before the Subcommittee on Africa, Global Human Rights and International Operations and the Subcommittee on Asia and the Pacific of the Committee on International Relations, February 15, 2006.
[20] "Isaac Mao and Michael Anti at Hong Kong U.," April 17, 2007, http://rconversation. blogs. com/ rconversation/2007/ 04/isaac_mao_and_m.html.
[21] Peter S. Goodman, "Yahoo Says it gave China Internet Data," *Washington Post*, September 11, 2005.
[22] Christine Chiao, "Microsoft Erases Anti-Blog," *AsiaMedia*, January 17, 2006.
[23] Google's Chinese service, with roughly 80 million customers and 30 million Gmail accounts, has captured 20%-30% of the PRC market, compared to *Baidu*, which has over 60%. Tom Krazit, "Google's Censorship Struggles Continue in China," *news.cnet.com*, June 16, 2009; Steven Mufson, "China Faces Backlash from 'Netizens' if Google Leaves," *Washington Post*, January 13, 2010.
[24] Robert McMillan, "Google Moving Search Records Out of China," *InfoWorld*, March 1, 2006; Rory Cellan-Jones, "China and Google: What's Going On," *BBC – Dot.Life*, June 25, 2009; James Mulvenon, "The Rule of Law in China: Incremental Progress," *The China Balance Sheet in 2007 and Beyond*, Center for Strategic and International Studies, May 2007.
[25] Claudine Beaumont, "China Accuses Google of Spreading 'Vulgar Content,'" *Telegraph.co.uk*, June 25, 2009.
[26] "Google Apologizes to Chinese Writers," *Agence France Presse*, January 11, 2010.
[27] Jonathan Mirsky, "China's Tyranny Has the Best Hi-Tech Help Censoring the Internet," *International Herald Tribune*, January 16, 2006.
[28] Steven Mufson, "China Turning to Technology to Hold onto Power," *Washington Post*, April 16, 2006; U.S. Congress, "The Internet in China: A Tool for Freedom or Suppression?" op. cit.
[29] In January 2010, a U.S. software firm filed a lawsuit against the Chinese government for copyright infringement, unfair competition, and other legal violations in connection with the Green Dam program. Agence France-Presse, "U.S. Software Firm Sues Chinese Government for US$2.2 Billion," *South China Morning Post*, January 6, 2010.
[30] "Green Dam Launch 'Not Handled Well'," http://www.chinaview.cn, August 14, 2009.

[31] Rebecca MacKinnon, "China Tightens Internet Controls in the Name of Fighting Porn, Piracy, and Cybercrime," *Rconversation* , December 14, 2009, http://rconversation.blogs.com; Sharon LaFraniere, "China Imposes New Internet Controls," *New York Times*, December 18, 2009.

[32] *Rconversation*, October 10, 2009, http://rconversation.blogs.com/rconversation/china/index.html.

[33] Estimates of the number of U.S. information technology, finance, defense, and other companies targeted in this attack ranged from 20 to 34.

[34] Google representatives stated that two Gmail accounts appeared to have been accessed but that the content of e-mail communications had not been breached. "Statement from Google: A New Approach to China," *Washington Post*, January 12, 2010. See also "A New Approach to China," *The Official Google Blog*, January 12, 2010, http://googleblog.blogspot.com/2010/01/new-approach-to-china.html.

[35] Jessica E. Vascellaro and Aaron Back, "Fallout from Cyber Attack Spreads—Google Investigates China Employees; Rift Emerges Between Yahoo! and Alibaba," *Wall Street Journal*, January 19, 2010; Rebecca MacKinnon, "Google Puts Its Foot Down," *RConversation,* http://rconversation.blogs.com/rconversation/china/index.html, January 13, 2010.

[36] Miguel Helft, "For Google, A Threat to China with Little Revenue at Stake," *New York Times*, January 15, 2010.

[37] "Clinton Urges Global Internet Freedom," *VOA News.com*, January 21, 2010; Gillian Wong, "China Denies Involvement in Google Hackings," *Washington Post*, January 25, 2010; "China Says Google 'No Exception to Law'," Embassy of the People's Republic of China in the United States, January 19, 2010.

[38] Prepared by Casey Addis, Analyst in Middle Eastern Affairs, 7-0846.

[39] "ITU Internet Indicators 2008," International Telecommunications Union, http://www.itu.int/ITU-D/icteye/Reporting/ShowReportFrame.aspx?ReportName=/WTI/InformationTechnologyPublic&RP_intYear=2008&RP_intLanguageID=1.

[40] "Speed Reduced for High Speed Internet in Iran," *BBC Persian*, October 20, 2006.

[41] See the U.S. State Department "2008 Human Rights Report: Iran," http://www.state.gov/g/drl/rls/hrrpt/2008/nea/119115.htm.

[42] NSN is a joint venture between the Finnish cell phone maker Nokia and the German company Siemens.

[43] Stuart Smith, "Politics of Marketing: Why Brands Continue to Surf the Recession," *Marketing Week* (London), August 13, 2009.

[44] See Christopher Rhoads and Loretta Chao, "Iran's Web Spying Aided by Western Technology," *Wall Street Journal*, June 22, 2009.

[45] Nokia Siemens Networks, "Provision of Lawful Intercept Capability in Iran," June 22, 2009, http://www.nokiasiemensnetworks.com/press/press-releases/provision-lawful-intercept-capability-iran.

[46] Eli Lake, "Fed Contractor, Cell Phone Maker Sold Spy System to Iran," *Washington Times*, April 13, 2009.

[47] Prepared by Gina Stevens, Legislative Attorney, 7-2581.

[48] Deep Packet Inspection ("DPI") is a computer network packet filtering technique that involves the inspection of the contents of data packets as they are transmitted across the network.

[49] Many foreign countries have privacy laws that may be applicable to Internet Service Providers, websites, etc. See Morrison & Foerster's Privacy Library for the text of privacy laws in other countries, in the U.S., and for multinational organizations, http://www.mofoprivacy.com/default.aspx?tabNum=2.

[50] See Andrew Noyes, "Chinese Demands for Web Filtering Software Cause a Stir," *CongressDailyAM*, June 25, 2009; Tim Wu, "The World Trade Law of Censorship and Internet Filtering," *Chi. J. Int'l L.*, vol. 7 (2006-07).

[51] Prepared by Kennon H. Nakamura, Analyst in Foreign Affairs, 7-9514.

[52] Under Secretary of State for Democracy and Global Affairs Paula Dobriansky, AOn-The-Record Briefing on the State Department=s *2006 Country Reports on Human Rights Practices,@* Washington, March 6, 2007, http://www.state.gov/g/ drl/rls/rm/2007/81468.htm.

[53] Ibid.

[54] U.S. Mission to the United Nations in Geneva, "Secretary of State Establishes New Global Internet Freedom Task Force," press release, February 14, 2006, http://geneva.usmission.gov/Press2006/ 021411nternet TaskForce.html.

[55] Secretary of State Hillary Rodham Clinton, "Remarks on Internet Freedom," January 21, 2010, http://www.state.gov/ secretary/rm/2010/01/135519.htm.

[56] On January 6, 1941, President Franklin Roosevelt addressed Congress saying that "we look forward to a world founded upon four essential human freedoms." These essential freedoms, which he referred to as the "Four Freedoms" are (1) freedom or speech and expression, (2) freedom of religion, (3) freedom from want, and (4) freedom from fear.

[57] The GIFT Strategy is available online at http://2001-2009.state.gov/g/drl/rls/78340.htm.

[58] Hillary Rodham Clinton, "Remarks on Internet Freedom," op. cit.

[59] Assistant Secretary of State for Democracy, Human Rights, and Labor Michael H. Posner, "Briefing on Internet Freedom and 21st Century Statecraft," January 22, 2010, http://www.state.gov/g/drl/rls/rm/2010/134306.htm.

[60] Questions following Secretary of State Hillary Clinton's *Remarks on Internet Freedom*, January 21, 2010, http://www.state.gov/secretary/rm/2010/01/135519.htm, and questions following Assistant Secretary of State

Michael Posner's "Briefing on Internet Freedom and 21st Century Statecraft," January 22, 2010, http://it.tmcnet.com/news/2010/ 01/26/4590599.htm.

[61] Originally prepared by Gina Stevens, Legislative Attorney, 7-2581.

[62] See http://www.globalnetworkinitiative.org/.

[63] Legislative summaries are taken directly from the Legislative Information Service of the Library of Congress.

[64] This bill is a substitute for H.R. 4595.

[65] Prepared by Patricia Moloney Figliola, Specialist in Telecommunication and Internet Policy, 7-2508. Adapted from "The State of Iranian Communication: Manipulation and Circumvention," Morgan Sennhauser, Nedanet, July 2009, http://iranarchive.openmsl.net/SoIC-1.21.pdf; and "Five Technologies Iran is Using to Censor the Web," Brad Reed, Network World, July 2009, http://www.networkworld.com/news/2009/072009-iran-censorship-tools.html.

[66] Adapted from *Reporters Without Borders*, "Handbook for Bloggers and Cyber-Dissidents," September 2005, http://www.rsf.org/IMG/pdf/Bloggers_Handbook2.pdf; and *The Citizen Lab*, "Everyone's Guide to By-Passing Internet Censorship for Citizens Worldwide," University of Toronto, September 2007, http://citizenlab.org/Circ_guide.pdf.

In: Technology Policies and Issues for the U.S. Government ISBN: 978-1-61209-501-1
Editors: Danielle A. Godden and Mason C. Andrews

Chapter 3

BIOTECHNOLOGY IN ANIMAL AGRICULTURE: STATUS AND CURRENT ISSUES*

Tadlock Cowan

SUMMARY

Animal agriculture is being transformed by rapid advances in biotechnology—a term that encompasses a variety of technologies, including genetic engineering (GE), genetic modification, transgenics, recombinant DNA techniques, and cloning, among others. Producers are interested in the application of biotechnology to improve productivity, consistency, and quality; to introduce new food, fiber, and medical products; and to protect the environment. Potential human health applications of transgenic animals include producing biopharmaceuticals and generating organs, tissues, and cells for xenotransplantation. Criticisms of such applications involve issues ranging from food safety and social resistance to potential negative impacts on animal welfare and on ecosystems. Questions also have arisen about the adequacy of the current regulatory structure to assess and manage any risks created by these technologies.

On January 15, 2009, the U.S. Food and Drug Administration (FDA) released final guidance on how it is to regulate GE animals and products. FDA is to do so under its existing statutory authority and regulations. Generally, GE-derived foods, for example, are to be regulated like non-GE foods; if their composition does not differ from their conventional counterparts, they will not have to be labeled. Nonetheless, developers of GE animals and of GE-derived products must gain FDA pre-market approval.

On February 6, 2009, the FDA announced the first approval of a drug from a GE animal. The drug is a human anti-clotting agent produced in the milk of transgenic goats. The FDA is currently considering approval of the first genetically modified animal for human

---

* This is an edited, reformatted and augmented version of Congressional Research Service publication, Report RL33334, dated September 10, 2010.

consumption, having declared in August 2010 that a GE salmon is safe to eat and poses no threat to the environment.

Although animal biotechnology involves many techniques other than cloning, this latter technology has attracted widespread attention. A final risk assessment and industry guidance on the safety of meat and milk from cloned cattle, pigs, and goats and their offspring were released January 15, 2008, by FDA. The documents generally echoed the FDA's December 28, 2006, draft risk assessment, which found that such products are as safe to eat as those of conventionally bred animals. The FDA also concluded that cloning poses the same risks to animal health as those found in animals created through other assisted reproductive technologies—although the frequency of such problems is higher in cloning. (Scientists stress that cloning is an assisted reproduction technique that does not involve any transfer or alteration of genes through GE.) The agency said it was no longer asking industry to refrain voluntarily from marketing the products of cloned animals and their offspring, although the U.S. Department of Agriculture (USDA) did ask that it be continued for products from clones (but not from the offspring of clones).

Bills on animal cloning introduced in the 110th Congress would have required all food from cloned animals or their offspring to be labeled, and prohibited food from cloned animals from being labeled as organic. These and other bills on cloning or other regulation of animal biotechnology could be offered in the 111th Congress.

INTRODUCTION

The U.S. Food and Drug Administration (FDA) released two recent documents that renewed public and congressional interest in animal biotechnology in general and animal cloning in particular. On January 15, 2009, the agency released final guidance representing its current thinking on the regulation of genetically engineered (GE) animals for food or drugs. FDA did so under its existing statutory authority and regulations.[1] A year earlier, on January 15, 2008, FDA had unveiled its final risk assessment and industry guidance on the safety of milk and meat from cloned animals and their offspring. The FDA found that, generally, meat and milk products from cattle, pigs, and goats are as safe as products from their non-cloned counterparts.[2]

Still, some in the United States remain concerned about the safety and other impacts of GE animals and of animal cloning, and they continue to advocate a cautious approach to commercialization of such products. Outside of the United States, animal biotechnology, including both GE and cloning, is the focus of ongoing regulatory, policy and scientific discussions within the European Union (EU) as well as within the international organization for animal health, known by its French acronym, the OIE. (Cloning by itself is not considered to be genetic engineering; see discussion later in this report.)

Biotechnology is a broadly defined term of relatively recent origin describing the range of modern knowledge, applications, and techniques underlying advances in many fields, notably health care and agriculture. Animal biotechnology has been defined as "that set of techniques by which living creatures are modified for the benefit of humans and other animals."[3] By its very nature, agricultural development is the history of humans modifying plants and animals to maximize desirable traits. For example, domestication and selective breeding of animals date back many thousands of years. Artificial insemination of livestock, notably dairy cattle, is a more recent technology, first finding wide commercial acceptance in the 1950s.

Discovery of the genetic code in the 1950s gave birth to modern techniques of biotechnology. One of the first commercial products of this new biotechnology in animal agriculture was bovine somatotropin (bST), a naturally occurring metabolic modifier that is now being manufactured in larger quantities through the use of recombinant DNA technology. Manufactured bST came onto the market in 1994 and is now administered to as many as half of all U.S. dairy cattle to increase per-cow milk output. Although bST is being used commercially in approximately 20 countries, it is banned in the European Union (EU).

The first U.S. approval of a commercial product from a GE animal is for the drug ATryn, an anti-clotting agent being produced in the milk of transgenic goats. The FDA announced its approval on February 6, 2009.

Other developments include pigs that have been engineered for increased sow milk output to produce faster-growing piglets. Cloned cattle also have been developed to resist mastitis, and transgenic salmon with enhanced growth characteristics are under regulatory consideration for possible commercialization. Output traits such as drugs recovered from animal milk ("pharming"), milk that lacks allergenic proteins, and animal organs for human transplant (xenotransplantation) that resist rejection are other contemporary objectives of animal biotechnology research. In March 2006, researchers at the University of Missouri announced the creation of transgenic pigs whose tissue contains omega-3 fatty acids.[4] The consumption of omega-3 fatty acids, found primarily in fish, has been linked to lowered incidence of heart disease in humans. Similar research is also under way to produce omega-3 fatty acids in cow's milk and in chicken eggs.

This report describes several scientifically emerging animal biotechnologies that are raising a variety of questions concerning risks to humans, animals, and the environment, as well as ethical concerns. The report examines applications of the technologies and discusses major issues that may arise. Consumers, agricultural producers, the biotechnology industry, and federal regulatory bodies are debating the relative costs and benefits of these technologies. As technologies move toward commercialization, Congress is being asked to examine these issues and possibly to refine the current federal regulatory structure governing the technologies and their agricultural products.

Animal Biotechnologies

Given the breadth of the term "animal biotechnology," one might reasonably define it to include thousands of years of humans selectively breeding animals: observing desirable animal traits and attempting to breed those traits into successive lines of animals. One of the first modern forms of assisted reproductive technology (ART) was artificial insemination (AI). AI has been long established as a technological advance in traditional selective breeding and an important adjunct to the development of modern industrial animal production, especially in dairy and poultry. AI was adopted by producers and accepted by the public with virtually no controversy. For example, more than 70% of all U.S.-bred Holstein cows, by far the most widely used milk producers, are artificially inseminated.[5] Estrus synchronization, which improves the efficiency of AI by more accurately controlling when a female is in heat, is also an important animal biotechnology.

With the development in the 1970s and patenting in the 1980s of recombinant DNA techniques, and the subsequent analysis of genes, their resulting proteins, and the role played by the proteins in animal biochemical processes (functional genomics), modern biotechnology is increasingly equipped with a set of sophisticated tools holding the promise of transforming the selective breeding of animals. The range of new techniques and technologies could transform animal biotechnology in ways that plant biotechnology was transformed in the 1980s and 1990s.

Modern animal biotechnology is developing against the background of public experience with plant biotechnology, and controversy over the technologies may be a continuing feature of animal biotechnology development, not least because of the closer connection between humans and some animals and the belief that techniques developed for animals are only a step away from application to humans. Some of the better known animal biotechnologies follow. A number of them are types of assisted reproductive technology (ART).

Embryo Transfer

After AI and estrus synchronization, embryo transfer (ET) is the third-most commonly used biotechnology. In ET, a donor cow of superior breeding is chemically induced to superovulate. The eggs are then fertilized within the donor, the embryo develops and is then removed and implanted in a recipient cow. Between removal and implantation, embryos may be frozen for safekeeping. Because of the relatively high costs, ET is used mostly within registered cowherds.

Transgenics

A prominent area of contemporary animal biotechnology research is the development of *transgenic animals* through genetic engineering (GE) technology.[6] Transgenic animals are produced by introducing an isolated DNA fragment into an embryo so that the resulting animal will express a desired trait. Transgenic animals may be generated by the introduction of foreign DNA obtained through animals of the same species, animals of different species, microbes, humans, cells, and *in vitro* nucleic acid synthesis. The only currently routine use of transgenic animals, primarily mice, is in the area of human disease research. The anti-clotting agent in goat milk is the first such application to be approved by the FDA (March 2009). Potential agricultural applications for such genetic engineering, however, include improved feed use and faster growth; more resistance to disease; meat that is leaner or that has more of some other desirable quality; and possibly even animal waste that is more environmentally benign. **Table 1** provides examples of various objectives of animal biotechnology involving genetic modification.

In Vitro Fertilization

With in vitro fertilization (IVF), a technician removes unfertilized eggs (oocytes) from the donor cow's ovaries, usually recovering 6-8 useable oocytes. The oocytes mature in an incubator and are fertilized with sperm. The resulting zygotes incubate and develop in the laboratory before being placed into the recipient cow. While IVF can produce many fertilized embryos, the added expense of ET makes the procedure prohibitive in most cases.

Sexing Embryos

The dairy industry prefers heifers and the beef industry prefers bulls. Embryo sexing methods in cattle have been developed using a bovine Y-chromosome probe. Technicians remove a few cells from the embryo and assess the DNA in these cells for the presence of a Y-chromosome. Presence of a Y-chromosome determines the embryo is male. Research is also developing in sperm sexing technology.

Table 1. Agricultural Applications of Animal Transgenics

Purpose	Animal model	Transgenic source
Faster growth/leaner meat	Cattle, swine, rabbits, sheep	Growth hormones/factors: *Human, Bovine, Porcine, Rat, Chicken*
Altered milk composition (higher protein)	Cattle	Extra copies of casein genes; disruption of lactoglobulin gene: *Cow*
Anti-clotting drug production in milk[a]	Goat	Human antithrombin gene
"Biosteel" production in milk[b]	Goat	Spider
Reduced phosphorus in swine feces	Swine	Phytase gene; *Bacteria*
Increased wool production	Sheep	Cysteine synthesis gene: *Bacteria* Growth factor: *Sheep*
Disease resistance	Swine, sheep, rabbit	Monoclonal antibodies: *Mouse* Viral envelope genes: *Sheep*
Xenotransplantation: Developing animal organs for human transplantation	Swine	CD55 (DAF-decay activating factor: *Human* CD59: *Human*

Source: GeneWatch UK, April 2002.

a. FDA approval for commercial use announced February 6, 2009. Source: FDA.

b. "Biosteel" is the trade name for spider web material intended to be produced in the milk of a transgenic goat. Said to be twenty times stronger than steel, "Biosteel" has an envisioned breaking strength of about 300,000 pounds per square inch and could produce microscopically fine, super strong fibers for industrial use.

Cloning

Cloning, discussed at greater length later in this report, is a biotechnology developing rapidly and with significant public controversy. Most people think of cloning as the creation of an organism that is genetically identical to another one. However, scientists use the term more broadly, to refer to production not only of such organisms but also of genetically identical cells, and to replication of DNA and other molecules. It also refers to a form of reproduction found naturally in many single-celled organisms, as well as plants and animals. Those differences in meaning and usage have caused some confusion in public debate about cloning, where the main area of controversy relates to artificial cloning involving higher organisms, including humans.[7]

Gene Knockout

This is a technique where researchers inactivate, or "knock out," a gene by replacing it or disrupting it with an artificial piece of DNA in order to determine what that particular gene does—for example, cause or protect against some disease, alter metabolism, and so forth. A knockout mouse is a laboratory mouse subjected to this technology.[8]

Regulation and Oversight

The basic federal guidance for regulating the products of agricultural biotechnology is the Coordinated Framework for Regulation of Biotechnology (51 *Fed. Reg.* 23302), published in 1986 by the White House Office of Science and Technology Policy (OSTP). A key principle has been that GE products should continue to be regulated according to their characteristics and unique features, not their methods of production, that is, whether or not they were created through biotechnology. The framework provides a regulatory approach intended to ensure the safety of biotechnology research and products, using existing statutory authority and previous policy experience.

Some newer applications of biotechnology did not exist when the current regulatory framework was enunciated. The NRC animal biotechnology report concluded that this regulatory regime "might not be adequate to address unique problems and characteristics associated with animal biotechnologies" and that federal agency responsibilities are not clear.[9]

Food and Drug Administration (FDA)

Within the Department of Health and Human Services (HHS), the FDA regulates food, animal feed ingredients, and human and animal drugs, primarily under the Federal Food, Drug, and Cosmetic Act (FFDCA; 21 U.S.C. §301 *et seq.*). The FDA has stated that most—although probably not all—gene-based modifications of animals for production or therapeutic claims fall within the purview of the agency's Center for Veterinary Medicine (CVM), which regulates them under the FFDCA as new animal drugs. A new animal drug (NAD) must be approved by the agency after it is demonstrated to be safe to man and animals, as well as being effective. Regulation of transgenic animals as NADs, however, suggests to some observers (e.g., the Center for Food Safety, Union of Concerned Scientists) the inherent weakness of existing regulatory structures to respond adequately to the complexities that arise with animal biotechnology innovations.[10]

Primarily under the FFDCA, FDA's Center for Food Safety and Applied Nutrition (CFSAN) is responsible for assuring that domestic and imported foods are safe and properly labeled. Generally, FDA does not review new foods themselves for safety before they enter commerce but does have enforcement authority to act if it finds foods that are adulterated under the act. All food *additives*, whether or not introduced through biotechnology, must receive FDA safety approval before they can be sold; the exception to pre-market approval are those on a list the FDA has determined to be "generally recognized as safe" (GRAS).[11]

Sections of the FFDCA and of the Public Health Service Act (42 U.S.C. §262 *et seq.*) provide the authorities for FDA's Center for Drug Evaluation and Research and Center for Biologics Evaluation and Research to regulate the safety and effectiveness of human drugs and other medical products, including those produced by GM animals. Under these laws, the FDA requires pre-market review and licensing of such products, and requires that their production conditions ensure purity and potency.[12]

FDA Approval of Drug in Goat Milk

The FDA's approval of the drug ATryn was the first of a commercially available product from a GE animal. FDA announced its decision on February 6, 2009, and published the final rule as a new animal drug application (NADA) in the February 11, 2009, *Federal Register* (74 *Fed. Reg.* 6823). The drug is an anticoagulant for the treatment of human patients with hereditary antithrombin deficiency; it is derived from the milk of goats bred with the introduction of an rDNA construct directing the expression of the human gene antithrombin. The milk is being produced by a herd of GE dairy goats in Massachusetts, which, the agency stated, had not demonstrated any negative health problems over seven generations. Controls are in place to ensure that no milk or meat from the goats will enter the food or feed supply, according to FDA. Biotechnology industry officials expressed their support for the approval; a leading animal activist group, the Humane Society of the United States, countered that the action "seems to perpetuate the notion of [animals] being merely tools for human use rather than sentient creatures."

FDA Guidance on GE Animals

On January 15, 2009, FDA's CVM released its industry guidance on how it plans to regulate GE animals.[13] This final document hews closely (with a few modifications) to the draft version that FDA published on September 18, 2008. The final document asserts that FDA's authority to regulate GE animals comes under the new animal drug provisions of the FFDCA. A drug is defined by the act, in part, as "intended to affect the structure or any function in the body of man or other animals." Also, part of the FFDCA definition of "new animal drug" is one intended for use in animals that is not generally recognized as safe and effective for use under the conditions prescribed or recommended, and that has not been used to a material extent or for a material time.

"The rDNA construct in a GE animal that is intended to affect the structure or function of the body of the GE animal, regardless of the intended use of products that may be produced by the GE animal, meets the FFDCA drug definition," the guidance states (on page 5). A new animal drug is considered "unsafe" unless FDA has approved an application for that particular use, or it is for investigational use and subject to an exemption from the drug approval requirement (among a few other specified exemptions). Therefore, most developers will have to submit to the "Investigational New Animal Drug" (INAD) process at FDA prior to shipping any GE animals or to marketing any food or feed derived from GE animals. In other words, it is illegal to introduce food from a GE animal into the food supply that has not been approved by FDA. The guidance lays out the pre-market approval process including the information to be required of developers.

Under the guidance, FDA will examine both the direct toxicity (including allergenicity) potential of food from a GE animal as well as any indirect toxicity. Generally, food and feed will be considered safe if the composition of edible materials from the GE animal can be shown to be as safe as from a non-GE animal. The labeling requirements for GE-derived foods would be the same as for other foods: FDA has oversight over the labeling of seafood, dairy products, and whole shell eggs, and USDA over the labeling of most meat, poultry, and egg products (see below). More specifically, food from GE animals would not have to be so labeled, except when it takes on a different character from its non-GE counterpart.

FDA Approval of Genetically Engineered Salmon

On August 25, 2010, the FDA announced that it had begun the approval process of a GE salmon—called AquAdvantage Atlantic Salmon—developed by the Massachusetts biotechnology firm AquaBounty. FDA also announced at the same time that it will hold a public comment period and a hearing on labeling for the transgenic salmon. Agency officials are undecided as to whether they would require any product labeling.

This would be the first genetically engineered animal approved for human consumption and commercial-level farming. FDA scientists stated in a briefing document that the GE salmon is safe for human consumption and poses no risk to the environment. These findings will be presented on September 19 to FDA's Veterinary Medicine Advisory Committee (VMAC), which will address science-based issues. These meeting will be open to the public. The VMAC will then advise officials whether to approve the salmon. The GE salmon has been engineered with a gene from the ocean eelpout that permits the salmon to grow at approximately twice the rate of a traditional Atlantic salmon. The GE salmon also contains a growth hormone from the Chinook salmon.

FDA is evaluating the GE salmon under its authority to regulate new veterinary drugs because the recombinant DNA construct that is intended to change the fish meets the definition of a drug, as defined under the Federal Food, Drug, and Cosmetic Act. This means that much of the supporting data AquaBounty supplies to FDA is confidential. A coalition of 31 organizations and restaurant chefs is demanding that FDA deny approval. Various environmental organizations are concerned that the GE salmon could escape from fish farms and threaten the wild salmon population. AquaBounty, however, says it would encourage producers to grow the GE Atlantic salmon only at in-land fish farms.[14]

U.S. Department of Agriculture (USDA)

Several USDA agencies, operating under a number of statutory authorities, also have at least potential roles in the regulation of transgenic and cloned animals and their products. As several critical reviews have indicated, USDA has not had a clearly spelled out policy in this area, including whether it intends to exercise these authorities to regulate GE animals.[15] USDA's Animal and Plant Health Inspection Service (APHIS) earlier had expressed its intention to publish an advance notice of proposed rulemaking (ANPR) on GE animals, possibly in 2008.[16] Instead, in concert with the FDA's notice on its draft guidance, APHIS published, in the September 19, 2008, *Federal Register*, a request for information from the public and scientists on how GE animals might affect U.S. animal health. Comments were accepted until November 18, 2008, as they had been for the FDA draft guidance. Further APHIS action was pending in mid-January 2009.

APHIS has broad authority, under the Animal Health Protection Act (AHPA; 7 U.S.C. §8301 *et seq.*) to regulate animals and their movement to control the spread of diseases and pests to farm-raised animals. APHIS also administers the Viruses, Serums, Toxins, Antitoxins, and Analogous Products Act (21 U.S.C. §151-159), aimed at assuring the safety and effectiveness of animal vaccines and other biological products, including those of GM origin, and the Animal Welfare Act (7 U.S.C. §2131 *et seq.*), portions of which govern the humane treatment of several kinds of warm-blooded animals used in research (but generally not agricultural animals). Elsewhere at USDA, the Food Safety and Inspection Service (FSIS) is responsible for ensuring the safety and proper labeling of most food animals and meat and related products derived from them under the Federal Meat Inspection Act (21 U.S.C. §601 *et seq.*) and Poultry Products Inspection Act (21 U.S.C. §451 *et seq.*).

Other Authorities

Reports and studies have cited a number of other authorities and federal agencies that are or could be relevant. The National Environmental Policy Act (NEPA; 42 U.S.C. §4321 *et seq.*) requires federal agencies to consider the environmental impacts of their actions.[17] The Environmental Protection Agency derives its authority from, among other laws, the Federal Insecticide, Fungicide and Rodenticide Act (FIFRA; 21 U.S.C. §301 *et seq.*); pesticides derived from living organisms, including those of biotechnology, are within its purview. The Interior Department's Fish and Wildlife Service and the Commerce Department's National Marine Fisheries Service have also been cited.[18]

Cloning Policy Developments

In 1997, scientists at the Roslyn Institute in Scotland used nuclei from the mammary cells of an adult sheep to clone "Dolly." Such nuclear transfer (NT) techniques were first developed in amphibians in the 1950s. They were first used in sheep in 1986, with the production of clones using nuclei taken from sheep embryos. The significance of Dolly was that she was cloned from differentiated cell types obtained from an adult (called "somatic cell nuclear transfer" or SCNT), rather than undifferentiated cells from an embryo ("embryonic NT").

Cloning in animal agriculture is generally not applied in isolation from other biotechnologies such as genetic engineering. Scientists note that cloning does not require fertilization and is not, by itself, a form of genetic engineering, that is, altering, removing, or inserting genes into an animal's existing DNA. However, cloning can involve transgenic as well as non-transgenic cells.

SCNT is not yet a notably efficient technique ready for widespread commercial adoption. For example, only about 6% of the embryos transferred to recipient cows resulted in healthy, long-term surviving clones, according to a 2005 report.[19] The European Food Safety Authority (EFSA) recently reported that overall success rates vary by species, ranging from 0.5% to 5%.[20]

Success rates are said to be improving, however. As more efficient cloning technologies, which can overcome the range of cloning abnormalities that have resulted from SCNT, are introduced, they could provide new opportunities in human medicine, agriculture, and animal welfare. This is the focus of much of the current international animal biotechnology research.

The EFSA draft scientific opinion estimated the number of live clones worldwide in 2007 to be less than 4,000 cattle and 1,500 pigs, of which about 750 cattle and 10 pig clones were in the United States. EFSA reported that life span data were limited, with only a few reports on cattle of six to seven years of age and no data available in 2007 on the full natural life span of livestock clones generally.[21]

FDA Risk Assessment

The FDA in July 2001 began asking companies to refrain voluntarily from marketing the products of cloned animals and their offspring until it could fully assess the scientific information on their safety. It subsequently issued two draft risk assessments on the matter, culminating in the release of a final risk assessment—and industry guidance that effectively lifted its voluntary moratorium—on January 15, 2008.

What Is Cloning?

Cloning, or somatic cell nuclear transfer, is a process by which animals are reproduced asexually.... In cloning, a differentiated somatic cell (a non-germ line cell from an existing animal) is introduced to an oöcyte (a cell that is the immediate precursor of a mature egg) that has had its nucleus (and thus its genome) removed, and then, following some manipulations, is induced to start replicating. If all goes well, the dividing cell is implanted into a female animal (dam), continues to develop normally, and is delivered just as any newborn.

—*Animal Cloning: A Risk Assessment, p. 20*

What's the Difference between a Cloned and a GE Animal?

Clones covered by FDA's January 2008 risk assessment and guidance are "just clones"—that is, they are copies of individual conventionally-bred animals, and do not contain any rDNA constructs. What can be confusing is that an animal clone can be genetically engineered (i.e., have an rDNA construct introduced into it), and a GE animal can be reproduced by cloning. The September 2008 guidance covers GE animals, regardless of whether they were reproduced by cloning, but does not cover animal clones that do not contain an rDNA construct.

—*Q and A to Accompany FDA's September 18, 2008 Draft Guidance*

Some stakeholders, including several Members of Congress, wanted a continuation of the moratorium until more studies are completed on safety and other aspects of animal cloning. Their views were reflected in nonbinding language to accompany the omnibus spending measure for FY2008, which passed in late 2007, and binding language in the Senate version of the pending farm bill (H.R. 2419); see "Congressional Activity," at the end of this report, for details.

In an October 2003 draft risk assessment, the FDA had concluded that "the current weight of evidence suggests that there are no biological reasons ... to indicate that consumption of edible products from clones of cattle, pigs, sheep or goats poses a greater risk than consumption of those products from their non-clone counterparts."[22] However, shortly after the assessment's publication, many members of the FDA's Veterinary Medicine Advisory Committee stated that there were not enough data to fully understand any potential risks from this relatively new technology.[23]

On December 28, 2006, the FDA's CVM released another long-awaited risk assessment—also in draft form, along with a proposed risk management plan and draft guidance for industry—on the safety of animal clones and their offspring.[24] The lengthy risk assessment examined two important questions: the safety of food from cloned animals and their progeny, and the effects of the process on the health of these animals. FDA officials stressed that the risk assessment did not address any other issues, such as the social and ethical aspects of cloning or consumer acceptance of cloned animal products. The risk assessment also focused only on cloning from non-transgenic cells. This risk assessment is now final.[25]

The final risk assessment concluded—as had the December 26, 2006 draft—that the meat and milk of clones of adult cattle, pigs, and goats, and the meat and milk from the offspring of these clones, were as safe for human consumption as the food from conventionally bred animals. The FDA added that not enough data were available to reach the same conclusions about sheep clones (or other species) and recommended that they not yet be used for human food. The risk assessment arrived at these conclusions after analyzing physiological, anatomical, health, and behavioral data on the animals and evaluating available information on the chemical composition of their milk and meat. FDA said its final assessment was peer-reviewed by an independent panel of experts, who agreed with the findings. The final assessment also took into account many thousands of public comments and additional data that became available since the draft was prepared, the agency said.

The agency said it would not require any special measures (including labeling) relating to the use of food products or animal feed derived from cloned cattle, goats, and pigs because they are "no different from food derived from conventionally bred animals. Should a producer express a desire for voluntary labeling (e.g., 'this product is clone-free'), it will be considered on a case-by-case basis to ensure compliance with statutory requirements that labeling be truthful and not misleading."[26] The industry guidance states that products from the offspring of cloned animals of any species are suitable for food or feed consumption.

Upon release of the 2006 proposed assessment, FDA officials said they were continuing to ask livestock breeders and producers to keep food products from cloned animals and their offspring out of commerce. This moratorium was announced in July 2001 and remained in effect until the final guidance was released in January 2008. Although the FDA moratorium is no longer in effect, USDA has since been encouraging technology providers to maintain the voluntary moratorium, but for products from cloned animals only. Food products from cloned animals' offspring would not be similarly constrained. During a continuation of this moratorium, USDA said it would "work closely with stakeholders to ensure a smooth and seamless transition into the marketplace for these products."[27]

Officials emphasized that, at least initially, almost all cloning-related foods will come not from the clones themselves but rather from their sexually reproduced offspring. Cloned animals—like other "elite" breeding animals such as a prized bull—are too valuable to use for food production, except possibly when they reach the end of their productive lives. The FDA observed that it can cost $20,000 or more to produce one such animal.

One ongoing question has been whether, in fact, milk and meat from the offspring of clones are entering the food supply. Several press accounts have quoted farmers and others who say they have sold such offspring to be slaughtered for food, and that the total number nationwide may be in the hundreds or even thousands. On the other hand, those numbers would amount to a small fraction of the total number of U.S. livestock slaughtered (e.g., 34 million cattle and 109 million hogs in 2007).[28]

FDA's January 2008 risk assessment did conclude that some animals involved in cloning (notably cattle and sheep surrogate dams) and some clones are at greater health risk than conventional animals. While the types of animal health problems observed in cloned animals are no different than those found with other assisted reproductive technologies, these problems appear more frequently in cloning than in the other technologies. Such problems include late gestational complications in the surrogate mothers, and increased risk of mortality and morbidity in calf and lamb clones that are apparently caused mainly by large offspring syndrome (LOS). Swine and goat clones and their mothers do not appear to experience any additional cloning-related problems, FDA reported.

FDA found that livestock clones as a group tend to have more health problems and death rates at or right after birth. However, the risk assessment added that most animals surviving the neonatal period appear to grow and develop normally, and that no increased risk of adverse health effects have been reported in clones approaching reproductive maturity. The agency said that the technology was too new to draw any conclusions on the relative longevity of livestock clones.

European Views

Cloning currently is not in commercial use in Europe, nor is there any specific authorization procedure in the European Union for products from cloned animals.[29] The European Food Safety Authority (EFSA) on July 24, 2008, released its scientific opinion on the food safety, animal health and welfare, and environmental impacts of animals derived from cloning. The EFSA opinion was limited to cattle and pigs, but its findings generally do not appear to be substantially different from those in the FDA assessment. According to EFSA, "Based on current knowledge, and considering the fact that the primary DNA sequence is unchanged in clones, there is no indication that differences exist in terms of food safety between food products from healthy cattle and pig clones and their progeny, compared with those from healthy conventionally-bred animals." As with FDA, the EFSA conclusion is based on an examination of compositional and nutritional data, the probability for the presence of novel constituents, the health status of the animal, and available data on toxicity, allergenicity, and microbiology.

EFSA agreed that SCNT can be a successful reproductive technique, although death and disease rates of clones are significantly higher than those of conventionally reproduced animals. EFSA noted that surrogate dams have higher rates of failed pregnancies and other problems. "A significant proportion of clones, mainly within the juvenile period for bovines and perinatal period for pigs, has been found to be adversely affected, often severely and with fatal outcome. Most clones that survive the perinatal period are normal and healthy," however. Adverse health effects were not observed in the offspring of clones, although studies have not been conducted for their entire lifespans.[30]

EFSA also found no indication of new or additional environmental risks compared with conventionally bred animals.

The EFSA report findings are being used by the European Commission (EC), the European Parliament, and member states as they discuss their policies regarding animal cloning. There is disagreement over the issue among officials in the various European bodies. This was illustrated when the European Parliament in early September 2008 approved a resolution asking the EC to propose a comprehensive ban on cloning animals for food, out of animal health and welfare and ethical concerns.[31] However, at an initial discussion of the issue in early January 2009, EC commissioners indicated that there was time to consider a science-based risk assessment before deciding whether or not to ban cloning.[32]

In April 2009, Members of the European Parliament (MEPs) called for the EU to introduce a ban on the marketing of meat and milk from cloned animals. The EC wanted meat and milk from cloned animals to be included under "novel foods." A ban could be difficult to impose because an EFSA Opinion does not find a food safety risk in the meat or milk of cloned animals. To counter a possible challenge in the WTO, some MEPs have argued that the EU should use ethical grounds to justify a ban on food from cloned animals and their offspring. Some observers allege that WTO regulations could permit member-states to impose bans on ethical grounds.

Other Views

Critics, including a number of consumer advocacy and animal rights groups, oppose the use of cloned animals and their offspring for food. In October 2006, a coalition formally petitioned the FDA to impose a moratorium on producing foods from cloned animals, and to establish rules for mandatory pre-market review and approval of cloned foods by regulating clones as new drugs under the food and drug act. These consumer and animal welfare groups contend that FDA has ignored or minimized a number of food safety and animal welfare problems. These include the treatment of surrogate mothers with high doses of hormones and their clone offspring, who often have severely compromised immune systems, with large doses of antibiotics, which could enter the human food supply; imbalances in clones' hormone, protein, or fat levels that could compromise meat and milk safety and quality; the possibility of increased foodborne illnesses; and a wide variety of health problems and abnormalities in the animals themselves. Consumer advocates assert that there are no consumer benefits from cloning.

FDA disagrees, citing the potential to breed livestock that meet consumers' changing tastes for traits like leanness, tenderness, color, size of meat cuts, and so forth. These would be in addition to potential producer-related benefits such as disease resistance, climate suitability, fertility, and improved physical qualities, FDA states.[33]

Opinion polls have suggested that the general public may not yet be ready for widespread animal cloning. A 2006 poll commissioned by the Pew Initiative on Food and Biotechnology, for example, found that Americans generally were not well informed about animal cloning, but 64% of those questioned are "uncomfortable" with it. Forty-three percent of Americans believed that foods from clones are unsafe, the survey found. Cloning opponents cite these types of findings to argue that products from cloned animals should be labeled so that consumers could avoid them if they wanted to. Some critics have urged the President to halt all FDA actions on cloning until a national panel that includes ethicists and religious leaders can consider the ethical and other social issues that they believe the technology has raised.

The NRC animal biotechnology report had stated that embryonic splitting and nuclear transfer using embryonic (not adult) cells were performed with some dairy cows to successfully produce genetically valuable offspring that were milked commercially and whose milk and meat did enter the food supply. Few concerns were raised by NRC authors about using these types of cloned animals for food, since they are generally believed to pose a low level of food safety concern. However, evaluating cloned-animal food composition "would be prudent to minimize any food safety concerns. The products of offspring of cloned animals were regarded as posing no food safety concern because they are the result of natural matings."[34] Other issues, notably consumer acceptance, social values, and animal welfare, could eventually overshadow any lingering questions about human health.

In August 2009, the Council for Agricultural Science and Technology (CAST), an international consortium of 33 professional and scientific societies based in Ames, IA, published a paper discussing issues surrounding animal cloning and transgenic animal research.[35] The report acknowledges that biotechnology proponents "have not convinced consumers that including these technologies in food production systems is in the consumer's best interest." The report also criticizes both FDA and USDA for failing to explain to the public the criteria for evaluating transgenic animals. Because FDA has taken a nontraditional approach to regulating transgenic animals by defining the transgene as a drug, the CAST authors argue that this "establishes an unusually high hurdle for the approval of a food product." The report further argues that a regulatory process in which consumers have confidence and with which companies can afford to comply must be in place for transgenic technologies to be applied to livestock.

Other Policy Concerns

The following are among the policy concerns that have arisen along with the development of new biotechnologies in animal agriculture. Some may be more applicable to GE-related technologies than to cloning *per se*, although others, like social acceptance and animal welfare concerns, may apply to both.

Environmental Issues

Environmental concerns arising from emerging animal biotechnologies are largely speculative at this time because few products have been commercialized. For example, although the EFSA draft scientific opinion foresaw no environmental impact, it also noted that limited data were available on this aspect of animal cloning.

Industrial developers of agricultural biotechnology might argue that more efficient production of animal-based feeds could reduce the resources necessary to produce food and, thereby, reduce the environmental burden of animal production. Should the development and widespread adoption of the "EnviroPig" (tm), which produces less phosphorus in its waste, occur, it might be considered by some to be a positive environmental benefit of agricultural biotechnology.

The 2002 NRC animal biotechnology report noted potential negative environmental impacts of genetically altered animals. Escape, survival, and gene flow into wild populations were identified as major concerns. Of most concern to the NRC committee was the escape into the environment of "super" salmon that have been genetically modified for rapid growth, and the likelihood that they could then breed with wild populations in the environment.[36] Other genetically altered animals such as fish, insects, and shellfish could also potentially escape into natural environments and become feral, disrupt ecosystems, or introduce novel genes in a natural population.

The FDA guidance on GE animals notes that the agency will comply with requirements of the National Environmental Policy Act (NEPA). Environmental risks are likely to differ depending upon the animal and application. For example, the types of environmental concerns arising from a GE cow bred for resistance to mastitis will differ greatly from the concerns raised by a GE freshwater fish engineered to grow more rapidly.[37] Material to accompany the FDA's final guidance notes that "although the agency has extensive experience in environmental assessment, including for fish, other federal and state agencies have overlapping or complementary authority and expertise" that FDA intends to tap. It also promised to make the results of environmental reviews public.[38]

Food Safety

Unexpected and unintended compositional changes arise with all forms of plant and animal genetic modification, including GE, concluded the IOM-NRC report on genetically engineered foods.[39] The report added that, so far, no GE-related adverse human health effects have been documented. However, the report's authors cited "sizeable gaps" in the ability to identify compositional changes caused by all forms of genetic modification—whether GE or conventional—and their relevance for human health, and they recommended new approaches for assessing the safety of new foods both before and after they enter the market.

Previous research and experience with commercializing transgenic plants suggested that negative effects on human health were virtually nonexistent. While not asserting that genetically modified organisms necessarily generate health problems, more recently reported research in peer-reviewed scientific journals has suggested that GMOs may raise food safety concerns:

- Australian researchers have published an article explaining that the transfer from a bean to a pea gene that expresses an insecticide protein has resulted in antibody production in mice fed the transgenic pea. The antibody reaction is a marker of allergic reaction.[40]
- Italian researchers at the University of Urbino had previously shown that absorption of transgenic soy by mice induced modifications in the nuclei of their liver cells. Recent research showed that a return to non-transgenic soy made the observed differences disappear.[41]
- Norwegian scientists at the University of Tromso demonstrated that the catalyst 35S CaMV, an element of the genetic structures used to modify a plant, can provoke gene expression in cultured human cells. This catalyst was previously believed to operate in this way only in plants.[42]

In the NRC animal biotechnology report, experts observed that the scientific principles for assessing the safety of GE animals are "qualitatively the same" as for non-GE animals.[43] However, because GE can introduce new proteins into foods, the potential for allergenicity, bioactivity, and/or toxicity responses should be considered, they said. Others have remarked that animals genetically engineered for nonfood products like pharmaceuticals or replacement organs might be of concern if such animals entered *or* affected the food supply.

Consumer and Social Acceptance

Criteria for selecting desirable traits to be produced through transgenic animals will likely be based on the demand for specific commercial characteristics. Even if scientific evidence is convincing that GE and cloned animal products are safe and beneficial for human consumption or economically valuable to producers, other concerns may limit marketplace and consumer acceptance.

Polls in recent years in the United States indicate that public knowledge about food and biotechnology generally remains limited. In two 2005 surveys, approximately half of those surveyed expressed opposition to the use of biotechnology in the food supply.[44] More than half of those in a 2005 Pew-sponsored poll said they opposed research into genetically modified animals, although opposition declined with increased knowledge. Many Americans have heard about animal cloning; two-thirds expressed discomfort with it—more of them out of religious or ethical concerns than food safety concerns. A majority of respondents to the Pew survey believe that regulators should take into account ethical and moral considerations. (See also the 2006 Pew findings on cloning, under "Cloning Policy Developments," section on "Other Views.")

Consumers may be less willing to accept the practice of genetically modifying animals than plants, some have argued, observing that people relate differently to animals, which many recognize as sentient beings. Some observers have expressed the concern that cloning farm animals might lead more quickly to human and pet cloning, which those observers oppose. Others believe that modifying animals, for example, to save human lives through xenotransplantation or the production of some important drug, might be more acceptable than doing so simply to produce more or cheaper food.

Further, science alone cannot resolve ethical views that appear to vary widely:

> Some people, irrespective of the application of technology, consider genetic engineering of animals fundamentally unethical. Others, however, hold that the ethical significance of animal biotechnologies must derive from the risk and benefits to people, the animals, and/or the environment. Yet another view focuses on the right of humans to know what they are eating or how their food or pharmaceuticals are being produced and therefore labeling becomes an issue to be addressed.[45]

Food industry leaders appear sensitive to consumer unease with animal biotechnology in general and cloning in particular. Many companies are not yet ready to process and sell meat and milk from transgenic and cloned animals because, they believe, consumers may view the products as less safe than more conventionally produced food products. The food industry also does not want to be portrayed as overstepping any widely held ethical or moral concerns about the new technologies.[46]

Such observations led some skeptics of animal biotechnology to propose that the FDA not only consider the science and safety issues, but also these broader concerns. In the area of human reproductive health, for example, the FDA and other federal agencies have invoked particular moral arguments either to reinforce scientific arguments or to counterbalance scientific evidence. Others believe that the FDA should base its decisions only on scientific evidence, and perhaps some other body should be established to consider the ethical and cultural questions. As the NRC animal biotechnology report observed, regulatory decisions and enforcement involving animal biotechnology "are difficult in the absence of an ethical framework."[47]

Labeling

Some believe that segregating and labeling the products of biotechnology in agriculture, including meat and milk, would enable the consumer to choose whether or not to buy such products. Either segregating and labeling biotechnology products or failing to do so could contribute to public suspicion that these products are flawed or different in some negative way, which may lead to contradictory policy decisions.[48] As noted, the FDA guidance on GE animals does not require their products to be so labeled. Opponents of labeling argue that biotechnology products essentially are the same as more conventional products—and are subjected to the same rigorous safety standards—and therefore should be treated no differently in the marketplace. A study by USDA's Economic Research Service reported that consumers' willingness to pay for a food item declines when the food label indicates that it was produced with the aid of biotechnology.[49]

In December 2007, two leading U.S. livestock cloning companies announced they were creating a voluntary system for tracking cloned animals. The program, which they asserted "is designed to facilitate marketing claims," is to involve a national registry providing for the individual identification of each cloned animal, affidavits that owners will sign committing to proper marketing and disposal of such animals, and monetary deposits to be returned when such commitments are fulfilled. Consumer organizations immediately criticized the initiative, noting that it would not track the offspring of cloned animals and was not mandatory.[50]

However, fierce debate continues regarding even the *voluntary* labeling of products to denote that they are *not* derived from biotechnology—notably the use of so-called "rBST-free" labels on dairy products. Although FDA cleared rBST as safe for commercial use in 1994, and it has been widely adopted by U.S. dairy farmers, the substance continues to be viewed skeptically by some consumer advocates. They have convinced a number of dairy processors and retailers to label their products as free of added BST manufactured through the use of recombinant DNA technology. FDA guidance permits such labeling so long as it is not misleading, is presented in the proper context, and is adequately substantiated. For example, a firm cannot state that its product is "BST-free" because the substance occurs naturally in milk, or even simply that it is "rBST-free" because that could imply a compositional difference between naturally occurring and rBST, which there is not, FDA has ruled. But certain claims that no rBST was used could be acceptable, the agency stated.[51]

FDA has looked to the states to evaluate the acceptability of such labels, and a number of these states have challenged processors' use of them. Battles over such restrictions have been notable in Pennsylvania, Ohio, Utah, Kansas, and elsewhere.[52]

Meanwhile, USDA has reminded stakeholders that products from cloned animals are not eligible to be labeled as organic, under its National Organic Program (NOP). While cautioning that the organics program is a marketing, not a food safety, program, USDA also noted: "Cloning as a production method is incompatible with the Organic Foods Production Act and is prohibited under the NOP regulations."[53] However, the status of products from clone offspring is less clear. The Department noted that USDA's Agricultural Marketing Service, where the NOP is located, was preparing rulemaking to address the organic status of such offspring.[54]

Animal Welfare

Some aspects of gene transfer, and of cloning, have the potential to create infectious disease hazards and/or impaired reproduction. Looming large in the ethical debate are questions about whether genetic modifications, cloning, and other technologies stress animals unnecessarily, subject them to higher rates of disease and injury, and hasten death.[55] The NRC animal agriculture report noted, for example, that ruminants produced by *in vitro* culture or nuclear cell transfer methods tend to have higher birth weights and longer gestation periods than those produced by artificial insemination, creating potential calving problems. Nuclear transfer techniques to propagate genetic modifications may increase risks to the reproductive health and welfare of both the surrogate female animals and their transgenic offspring. The report cited other evidence of problems such as anatomical, physiological, or behavioral abnormalities in many transgenic animals.[56] Some scientists have countered that animal welfare problems have been exaggerated and tend to recede, particularly as the technologies are perfected. Most appear to agree, however, that animals originating from some forms of genetic modification or from cloning may require closer observation and care. (See also the section of this report on "Cloning Policy Developments.")

In Europe, officials continue to consider the ethical dimensions of animal welfare in formulating their own cloning policies. The EC charged a separate panel, the European Group on Ethics in Science and New Technologies, with drafting an opinion. This group released an opinion on January 16, 2008, stating, in part:

> Considering the current level of suffering and health problems of surrogate dams and animal clones, the EGE has doubts as to whether cloning animals for food supply is ethically justified. Whether this applies also to progeny is open to further scientific research. At present, the EGE does not see convincing arguments to justify the production of food from clones and their offspring.

If such products were to be allowed into the market, certain requirements regarding food safety, animal welfare and traceability should be met, the group added.[57]

Genetic Diversity

Could the introduction of a few genetically altered or cloned "superspecies" bring too much genetic uniformity to herds? As genetic diversity declines, herds could be more susceptible to diseases, leading to large production losses and/or much heavier use of antibiotics and other animal drugs to treat them, some have argued. A related concern is that a relative handful of "elite" producers or breeders might hold the proprietary rights to these species, to the disadvantage of many farmers and ranchers. Some animal biotechnology researchers have pointed to the potential importance of preserving unaltered germlines in domestic animals because they could prove to be an invaluable "gene bank" in the event that novel infectious diseases or inheritable genetic defects were inadvertently introduced into modified subpopulations as a consequence of genetic modification.

Trade Issues

If the United States were to be the first country to approve food products from cloned animals, how might the decision affect U.S. exports? Any exports of the products of animal biotechnology would presumably encounter a wide spectrum of foreign regulatory regimes, some more restrictive than the U.S. system. For example, the current European Union restriction on new biotechnology products is likely to encompass various restrictions on animal biotechnology as it does on plant biotechnology. On the other hand, researchers in a number of other countries, including some EU members, have been producing clones, and one—France—has published its own risk assessment on clones that, FDA has observed, generally agrees with the U.S. assessment.[58]

International guidelines pertaining to exports of animal products derived from biotechnology are being considered. The Codex Ad Hoc Intergovernmental Task Force on Foods Derived from Biotechnology held an initial meeting in September 2005 in Chiba, Japan, to determine the new work projects. Almost every country, except for the United States, proposed an animal biotechnology project. The task force agreed to move forward with a recombinant DNA (r-DNA) animal project, specifically to develop guidelines for how countries would assess the safety of foods derived from r-DNA animals.[59] At a meeting in November-December 2006 in Chiba, the task force, among other things, reviewed the developing guidelines, agreeing to limit their scope to food safety and nutritional issues (while recognizing the importance of others like animal welfare, environmental, and ethical concerns).[60]

USDA has said it would consider "discussion with industry on possible verification of its supply chain management plan to ensure that trading partners are aware of whether or not they receive cloned or non-cloned products."[61]

Imports of GE animals and their products into the United States also are of concern to some. In a December 2008 audit report, USDA's Office of Inspector General (OIG) concluded that the Department has not established an import control policy for such imports but needs to do so. "To mitigate any risks to the U.S. environment, agriculture, and commerce from unapproved transgenic plants and animals entering the U.S. food supply, USDA will need to monitor" foreign developments in such technologies more closely, OIG also noted.[62]

Congressional Activity

Members of Congress in the past have proposed various bills aimed at more closely regulating the plant and animal products of agricultural biotechnology generally, requiring such products to be labeled, and providing a means to recover any damages caused by the technology. As with human cloning, ethical issues concerning animal clones and other animal biotechnologies also may continue to be visible public issues. As before, the 111th Congress could be asked to play a larger role in weighing the benefits and costs of these evolving technologies, and to refine existing government oversight.

In the 110th Congress, the Senate-passed version of the 2007-2008 farm bill (H.R. 2419, in Section 7507) contained statutory language that would have required the FDA both to postpone publication of its final risk assessment until the completion of several newly mandated studies, and also to maintain the voluntary marketing moratorium until then. However, conferees deleted the Senate language from the final measure (signed into law as P.L. 110-234).

More specifically, the Senate bill would have directed the HHS Secretary to contract with the National Academy of Sciences to conduct a study on the safety of food products from cloned animals and the health effects and costs attributable to milk from cloned animals, with a report to Congress due within one year of enactment. The study was to address whether there were a sufficient number of studies to support the FDA draft assessment, and whether there were other pertinent ones that were not taken into account. It would have included an evaluation of potential public health effects and associated health care costs, and an evaluation of any consumer behavior and negative health and nutrition impacts resulting from a decrease in dairy consumption if milk from cloned animals and their offspring is commercialized.

Animal cloning-related bills (S. 414 and H.R. 992) were introduced in early 2007, both of which would have amended federal food safety laws to require that foods from cloned animals and their offspring be so labeled. Also introduced in early 2007 was S. 536/H.R. 1396, which would have prohibited the use of the "organic" label on food products from cloned livestock or their offspring. Another (H.R. 4855), introduced in late 2007, would have required studies like those in the Senate farm bill on the impacts of food products from cloned animals entering the food supply.

In late July 2008, three related bills were introduced that, their sponsor said, were to create a comprehensive framework for regulating GMOs. H.R. 6636 would have required the labeling of all foods produced with GE material. H.R. 6635 would have prescribed relatively stringent new regulations for FDA approval and oversight of GE crops. H.R. 6637 would have regulated business dealings between agricultural producers and the developers of genetically engineered plants and animals, and sought to hold biotechnology companies liable for any adverse on-farm impacts from their products.

These bills were not enacted, nor have they been reintroduced in the 111th Congress.

Acknowledgments

This report was originally written in collaboration with Geoffrey S. Becker, Specialist in Agricultural Policy.

End Notes

[1] CVM, FDA, Guidance for Industry: *Regulation of Genetically Engineered Animals Containing Heritable rDNA Constructs*. The guidance can be accessed through the FDA website at http://www.fda.gov/OHRMS/ DOCKETS/ 98fr/ FDA-2008-D-0394-gdl.pdf.

[2] FDA documents were posted on the Internet at http://www.fda.gov/cvm/cloning.htm on January 15, 2008.

[3] National Research Council (NRC), *Animal Biotechnology: Science-Based Concerns*, (Washington, DC, National Academy Press, 2002). (Hereafter cited as NRC, *Animal Biotechnology*.) Unless noted, this CRS report is based on material in that document and the following additional sources: Institute of Medicine (IOM) and NRC, *Safety of Genetically Engineered Foods: Approaches to Assessing Unintended Health Effects* (Washington, DC, National Academy Press, 2004). (Hereafter cited as IOM-NRC, *Safety of Genetically Engineered Food*.); Council for Agricultural Science and Technology (CAST), *Biotechnology in Animal Agriculture: An Overview* (Issue Paper 23) (Washington, DC, February 2003). (Hereafter cited as CAST, *Animal Agriculture*.); Pew Initiative on Food and Biotechnology, *Issues in the Regulation of Genetically Engineered Plants and Animals*, (Washington, DC, Pew Initiative April 2004. See http://pewagbiotech.org/research/regulation/.) (Hereafter cited as Pew, *Issues in Regulation*.); Pew Initiative on Food and Biotechnology, *Post-Market Oversight of Biotech Foods* (Washington, DC, Pew Initiative April 2003) (and other Pew materials); information accessed through the websites of the Biotechnology Industry

Organization at http://www.bio.org/; the Center for Food Safety, at http://www.centerforfoodsafety.org/; the Union of Concerned Scientists, at http://www.ucsusa.org/; and the Consumer Federation of America, at http://www.consumerfed.org/.

[4] *New York Times*, March 27, 2006.

[5] FDA, "A Primer on Cloning and Its Use in Livestock Operations," at http://www.fda.gov/cvm/ Cloning RA_Primer.htm.

[6] *Genetic engineering* (GE) here refers to the use of molecular biology to alter cells by inserting or removing genes. GE is a form of *genetic modification (GM)*, which refers more broadly to the practices of altering an organism's genetic composition by both GE and non-GE methods.

[7] See CRS Report RL31358, *Human Cloning*, by Judith A. Johnson and Erin D. Williams.

[8] National Institutes of Health, National Human Genome Research Institute, "Knockout Mice," accessed January 21, 2009, at http://www.genome.gov/12514551.

[9] NRC, *Animal Biotechnology*, p. 14.

[10] See Center for Food Safety website, at http://www.centerforfoodsafety.org/geneticall7.cfm; and the Union of Concerned Scientists website, at http://www.ucsusa.org/food_and_agriculture/. Also see discussion of FDA's guidance for industry, later in this section.

[11] The FDA, as of early 2009, has not granted approval for any human foods from transgenic (or cloned) animals, although a "very limited number have been approved for rendering into animal feed components." The only FDA-approved product of biotechnology in wide commercial use is bST, and an application is pending for a GE salmon. (*Questions and Answers about Transgenic Fish*, at http://www.fda.gov/cvm/transgen.htm). Also see CRS Report RL34247, *Federal Regulation of Substances Generally Recognized As Safe (GRAS) and the Use of Carbon Monoxide in Packaging for Meat and Fish*, by Vanessa K. Burrows and Cynthia Brougher.

[12] NRC, *Animal Biotechnology*, p. 163. At a January 9, 2009 meeting, the FDA Blood Products Advisory Committee ruled that a new drug to prevent clots in humans, recombinant human antithrombin III produced in the milk of GE goats, is safe and effective. The FDA could make a final regulatory decision to approve the license application of its developer, GTC Biotherapeutics, soon. Source: "FDA issues final guidance to industry on transgenic animals," *Food Chemical News*, January 19, 2009.

[13] *Guidance for Industry: Regulation of Genetically Engineered Animals Containing Heritable rDNA Constructs.* FDA noted that much of the new guidance will be relevant also to non-heritable rDNA constructs (such as modifications intended for gene therapy); a separate guidance for non-heritable constructs might come later. The agency states at the outset: "This guidance represents the Food and Drug Administration's (FDA's) current thinking on this topic. It does not create or confer any rights for or on any person and does not operate to bind FDA or the public. You can use an alternative approach if the approach satisfies the requirements of the applicable statutes and regulations."

[14] Research published in the *Proceedings of the National Academy of Sciences* notes that a release of just 60 GE salmon into a wild population of 60,000 would lead to the extinction of the wild population in less than 40 generations. W. M. Muir and R. D. Howard, "Possible ecological risks of transgenic organism release when transgenes affect mating success: Sexual selection and the Trojan gene hypothesis." *Proceedings of the National Academy of Sciences*, 96: 13853-13856 (1999).

[15] See, for example, Pew, *Issues in Regulation*. Beginning on p. 139, the report contains an extensive discussion on how these and several other USDA authorities might be utilized for oversight of animal biotechnology.

[16] APHIS in 2007 established an Animals Branch within its Biotechnology Regulatory Services "to develop a regulatory framework for the possible regulation of genetically engineered animals." Source: APHIS website, "Regulation of Genetically Engineered Animals," at http://www.aphis.usda.gov/biotechnology/ news_transgenic_animals.shtml.

[17] The FDA guidance discusses how NEPA requirements will apply to the GE animal approval process.

[18] The FDA guidance on GE animals states that the agency will work with other relevant federal and state agencies should it receive a request for approval of a GE animal intended for release into the wild.

[19] D. N. Wells, "Animal Cloning: Problems and Prospects," *Review of Science and Technology*, 24(1): 251-264, 2005.

[20] European Food Safety Authority. *Draft Scientific Opinion on Food Safety, Animal Health and Welfare and Environmental Impact of Animals Derived from Cloning by Somatic Cell Nucleus Transfer (SCNT) and their Offspring and Products Obtained from the Animals*, Enorsed for public consultation December 19, 2007 and released January 11, 2008. Hereinafter cited as EFSA draft scientific opinion. Accessed on the Internet at http://www.efsa.europa.eu/EFSA/ efsa_locale-1178620753812_1178676922939.htm.

[21] Ibid.

[22] See http://www.fda.gov/cvm/Documents/CLRAES.pdf.

[23] "Panel Calls for More Data on Animal Cloning Risks," *Food Chemical News*, November 10, 2003.

[24] Availability of the draft documents was formally published in the January 3, 2007, *Federal Register* (72 FR 135-137; FDA docket no. 2003N-0573); comments were accepted until May 3, 2007.

[25] FDA documents were posted on the Internet at http://www.fda.gov/cvm/cloning.htm on January 15, 2008. They are the Risk Assessment, Risk Management Plan, and Guidance for Industry.

[26] "FDA Issues Documents on the Safety of Food from Animal Clones," FDA press release, January 15, 2008.
[27] Statement by Bruce Knight, Under Secretary for Marketing and Regulatory Programs on FDA Risk Assessment on Animal Clones, January 15, 2008.
[28] See, for example, "Animal Clones' Offspring Are in Food Supply," *The Wall Street Journal* online, September 2, 2008.
[29] "EFSA launches its draft opinion on animal cloning for public consultation," EFSA press release, January 11, 2008.
[30] "Scientific Opinion of the Scientific Committee on a request from the European Commission on Food Safety, Animal Health and Welfare and Environmental Impact of Animals derived from Cloning by Somatic Cell Nucleus Transfer (SCNT) and their Offspring and Products Obtained from those Animals," *The EFSA Journal* (2008) 767, 1-49; http://www.efsa.europa.eu/EFSA/efsa_locale-1178620753812_1211902019540.htm.
[31] "European Parliament urges ban on cloning for food," *Food Chemical News*, September 8, 2008.
[32] "EU not ready to ban animal cloning for food," *Food Chemical News*, January 19, 2009.
[33] "A Primer on Cloning and Its Use in Livestock Operations," at http://www.fda.gov/cvm/CloningRA_Primer.htm.
[34] CAST, *Animal Agriculture*, p. 8.
[35] *Animal Productivity and Genetic Diversity: Cloned and Transgenic Animals. Issue Paper 43*, Council for Agricultural Science and Technology, August 2009. Available at http://www.cast-science.org/publications.asp.
[36] NRC, *Animal Biotechnology*, p. 73.
[37] FDA, *Consumers Q&A*, accompanying its September 18, 2008, draft guidance.
[38] "FDA's Response to Public Comments," accessed January 21, 2008, at http://www.fda.gov/cvm/ GEanimals. htm#Consumer%20Health%20Information.
[39] IOM-NRC, *Safety of Genetically Engineered Food*, p. 1.
[40] V. E. Prescott, P. M. Campbell, et al., "Transgenic Expression of Bean-Amylase Inhibitor in Peas Results in Altered Structure and Immunogenicity," *Journal of Agriculture and Food Chemistry*, 53(23); 9023-9030, November 2005.
[41] M. Malatesta, C. Tiberi, et al., "Reversibility of Hepatocyte Nuclear Modifications in Mice Fed on Genetically Modified Soybean," *European Journal of Histochemistry*, 49(2):237-242, July-September 2005.
[42] M. R. Myhre, K. Fenton, et al., "The 35S CaMV Plant Virus Promoter Is Active in Human Enterocyte-like Cells," *European Food Research and Technology*, 222(1-2):185-193, January 2006.
[43] NRC, *Animal Biotechnology*, p. 65.
[44] Gallup poll on biotechnology and food safety, July 2005; Mellman Group/Public Opinion Strategies Poll Conducted for the Pew Initiative on Food and Biotechnology, October 2005. See http://www. pewagbiotech.org (accessed March 20, 2006).
[45] NRC, *Animal Biotechnology,* p. 13.
[46] These concerns were explored in depth at an October 2006 symposium, "Animal Biotechnology: Considering Ethical Issues," sponsored by the Pew Initiative on Food and Biotechnology and by Michigan State University. Also reported by *Food Chemical News*, October 23, 2006.
[47] Ibid.
[48] C. T. Foreman, *Can U.S. Support for Food Biotechnology Be Salvaged?*, paper prepared for the American Enterprise Institute for Public Policy Research Conference on Biotechnology, the Media and Public Policy, June 12, 2003.
[49] A. Tegene, W. Huffman, et al., *The Effects of Information on Consumer Demand for Biotech Foods: Evidence from Experimental Auctions*, USDA, Economic Research Service, Technical Bulletin 1903, March 2003.
[50] Sources: "Top U.S. Cloning Companies Announce New System to Track Cloned, Livestock," December 19, 2007, Viagen press release, and "Consumer Groups Displeased with Voluntary Animal Clone Registry," *FDA Week*, December 21, 2007.
[51] FDA, "Interim Guidance on the Voluntary Labeling of Milk and Milk Products From Cows That Have Not Been Treated With Recombinant Bovine Somatotropin," *Federal Register*, February 10, 1994.
[52] See, for example, "Dairy processors fight to save 'rBST-free' labels, March 3, 2008, and "Kansas is latest battleground in 'rBST-free' labeling fight," December 22, 2008, both in *Food Chemical News*.
[53] USDA, Questions and Answers, FDA's Final Risk Assessment, Management Plan and Industry Guidance on Animal Clones and their Progeny, January 2008.
[54] Ibid. Such a rule was not unveiled in 2008.
[55] See also H. P. S. Kochhar, G. Adlakha-Hutcheon, and B. R. Evans, "Regulatory Considerations for Biotechnology-Derived Animals in Canada," *Review of Science and Technology*, 24(1):117-125, 2005.
[56] NRC, *Animal Biotechnology*, p. 11.
[57] "Ethical aspects of animal cloning for food supply," Opinion No. 23. The group's website is at http://ec.europa.eu/ european_group_ethics/index_en.htm.
[58] Stephen F. Sundlof, then Director of FDA Center for Veterinary Medicine, transcript of a December 28, 2006, teleconference on the draft risk assessment.

[59] Codex is recognized by the World Trade Organization (WTO) as the body that sets food safety standards for facilitating international trade of food products. The WTO cites Codex texts as a benchmark in the Agreement on Sanitary and Phytosanitary Measures (SPS).

[60] The collaborative document is entitled *Proposed Draft Guideline for the Conduct of Food Safety Assessment of Foods Derived from Recombinant-DNA Animals*. Codex adopted the guidelines at its Geneva meeting on June 30-July 5, 2008.

[61] Questions and Answers, FDA's Final Risk Assessment, Management Plan and Industry Guidance on Animal Clones and their Progeny.

[62] USDA, OIG, *United States Department of Agriculture Controls Over Importation of Transgenic Plants and Animals*, Audit Report No. 50601-17-Te, December 2008.

In: Technology Policies and Issues for the U.S. Government ISBN: 978-1-61209-501-1
Editors: Danielle A. Godden and Mason C. Andrews

Chapter 4

AGRICULTURAL BIOTECHNOLOGY: BACKGROUND AND RECENT ISSUES*

Tadlock Cowan

SUMMARY

U.S. soybean, cotton, and corn farmers have rapidly adopted genetically engineered (GE) varieties of these crops since their commercialization in the mid-1990s. Over the last decade, GE varieties in the United States have increased from 3.6 million acres to 143 million acres. Worldwide, 25 countries planted GE crops on approximately 309 million acres in 2008. GE varieties now dominate soybean, cotton, and corn production in the United States, and they continue to expand rapidly in other countries. As adoption has spread, policy debates have continued over the costs and benefits of GE products.

Ongoing concerns include the impacts of GE crops on the environment and food safety, and whether GE foods should be specially labeled. Underlying these issues is the question of whether U.S. regulation and oversight of biotechnology—with responsibilities spread primarily among the U.S. Department of Agriculture (USDA), the Food and Drug Administration (FDA), and the Environmental Protection Agency (EPA)—are adequate, particularly as newer applications, for example, biopharmaceuticals (drugs manufactured with the use of GE crops or animals) or stacked GE traits in single organisms, emerge that did not exist when the current regulatory regime was established.

Regulatory noncompliance incidents most pointedly raise concerns about the adequacy of existing U.S. regulatory structures. About 16 major events have occurred since 1995, according to USDA's Animal and Plant Health Inspection Service (APHIS). Another recurring concern has been the adequacy of APHIS's environmental assessments (EAs) for deregulating GE plants. In 2006, a U.S. district court held that USDA's EA for a variety of GE alfalfa was inadequate for issuing a fmding of no significant impact (FONSI); the court

* This is an edited, reformatted and augmented version of Congressional Research Service publication, Report RL32809, dated September 2, 2010.

vacated APHIS's decision to deregulate GE alfalfa and ordered APHIS to complete an environmental impact statement (EIS). The federal court subsequently enjoined further planting until the EIS was completed. In July 2010, the Supreme Court overturned the federal court's decision to enjoin planting. The final EIS is expected by the end of the year, but APHIS has not indicated an intent to allow planting under a "partial deregulation" in the interim. A similar case involves APHIS's decision in 2005 to deregulate GE sugar beets on the basis of its EA. That decision was also challenged, and the federal court vacated the deregulation of the GE sugar beets and ordered APHIS to complete an EIS. In July 2010, the federal court declined to enjoin GE sugar beet planting, but in vacating APHIS's decision to deregulate the GE beets, the court has effectively halted planting, except under regulated conditions. APHIS is currently evaluating a request to partially deregulate GE sugar beets for 2011.

In October 2008, APHIS announced the first revision of its biotechnology regulations since their promulgation in 1987. Proposed changes include a multi-tiered permitting system, new risk categorizations for assessing environmental releases of GE organisms, regulation of GE plants that produce pharmaceutical and industrial compounds, and new standards for low-level presence of regulated GE products. A final rule on the proposed changes has not yet been published. Other recent issuances include FDA's January 2009 fmal guidance on regulation of GE animals and products. APHIS is now seeking public comment and data concerning ongoing and future research on GE animals. In a ruling in January 2008, APHIS published its final guidance on the safety of meat and milk from cloned animals.

Legislative activity in the 111th Congress was modest. Two bills (H.R. 5578 and H.R. 5579) were introduced by Representative Kucinich, who had introduced similar bills in previous years.

Adoption of Biotechnology in Agriculture[1]

Farmers have always modified plants and animals to improve growth rates and yields, create varieties resistant to pests and diseases, and infuse special nutritional or handling characteristics. Such modifications have been achieved by crossbreeding plants and animals with desirable traits, through hybridization, and other methods. Now, using recombinant DNA techniques, scientists also genetically modify plants and animals by selecting individual genes that carry desirable traits (e.g., resistance to a pest or disease) from one organism, and inserting them into another, sometimes very different, organism, that can be raised for food, fiber, pharmaceutical, or industrial uses.

Karl Ereky, a Hungarian engineer, coined the term "biotechnology" in 1919 to refer to the science and the methods that permit products to be produced from raw materials with the aid of living organisms.[2] According to the Convention of Biological Diversity, biotechnology is "any technological application that uses biological systems, living organisms, or derivatives thereof, to make or modify products or processes for specific use" (Article 2). According to the FAO's statement on biotechnology, "interpreted in a narrow sense, ... [biotechnology] covers a range of different technologies such as gene manipulation and gene transfer, DNA typing and cloning of plants and animals."[3]

Since genetically engineered (GE, sometimes called genetically modified organism or GMO) crop varieties first became commercially available in the mid-1990s, U.S. soybean, cotton, and corn farmers have rapidly adopted them in order to lower production costs and increase crop yields. Proponents point to the emergence of "second generation" GE commodities that could shift the focus of biotechnology from the "input" side (creating traits that benefit crop production, such as pest resistance) to the "output" side (creating traits that benefit consumers, such as lower- fat oils). These second generation products could offer enhanced nutritional and processing qualities and also industrial and pharmaceutical uses. Future products are expected to be livestock- as well as crop-based. Critics, meanwhile, complain that biotechnology companies generally have not yet delivered the consumer benefits they have been promising for years.

Incidents of regulatory noncompliance have continued to spike concern about the adequacy of regulatory structures. In December 2008, a small amount of unapproved GE cotton was harvested along with commercially available GE cotton. The unapproved GE cotton variety produces a pesticide that is a plant-incorporated protectant (PIP). In August 2006, traces of an unapproved variety of GE rice were reported in commercial rice samples from parts of the southern United States (see "GE Rice," below). These incidents have added to the ongoing interest in a number of public policy questions. What are the environmental and food safety impacts of GE crops and animals? What obstacles and opportunities are exporters of GE crops encountering in the global marketplace? Is the current U.S. regulatory framework, which is based primarily upon statutory authorities enacted before the rise of agricultural biotechnology, adequate for these new technologies and products?

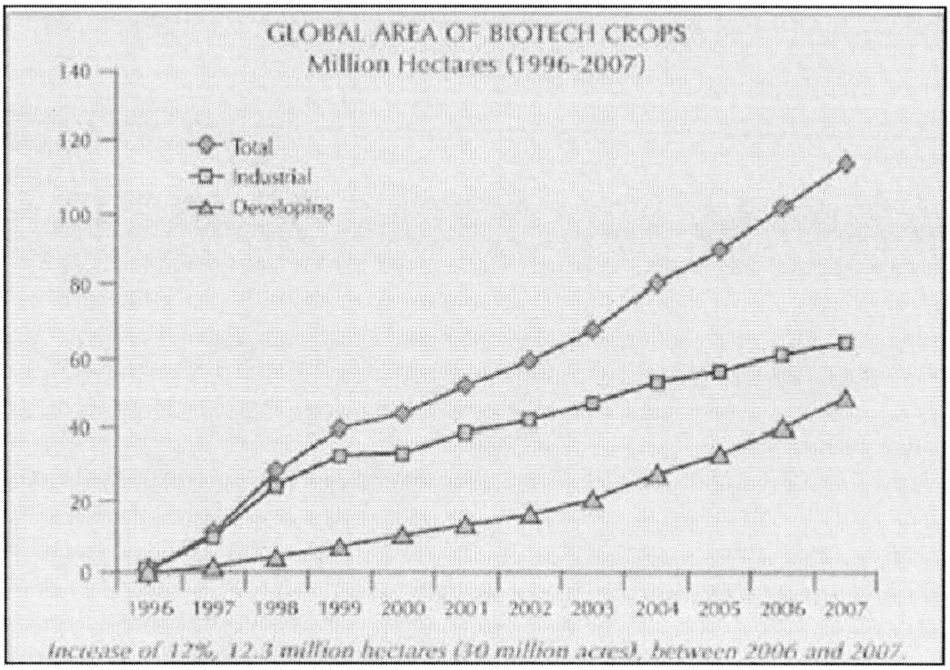

Source: International Service for the Acquisition of Agri-biotech Applications (ISAAA), Clive James, 2007 ISAAA Report on the Global Status of Biotech/GM Crops. Accessed July 2009 at http://www.isaaa.org/resources/ publications/briefs/37/executivesummary/ default. html.

Figure I. Global Acreage Planted to GE Crops, 1996-2009

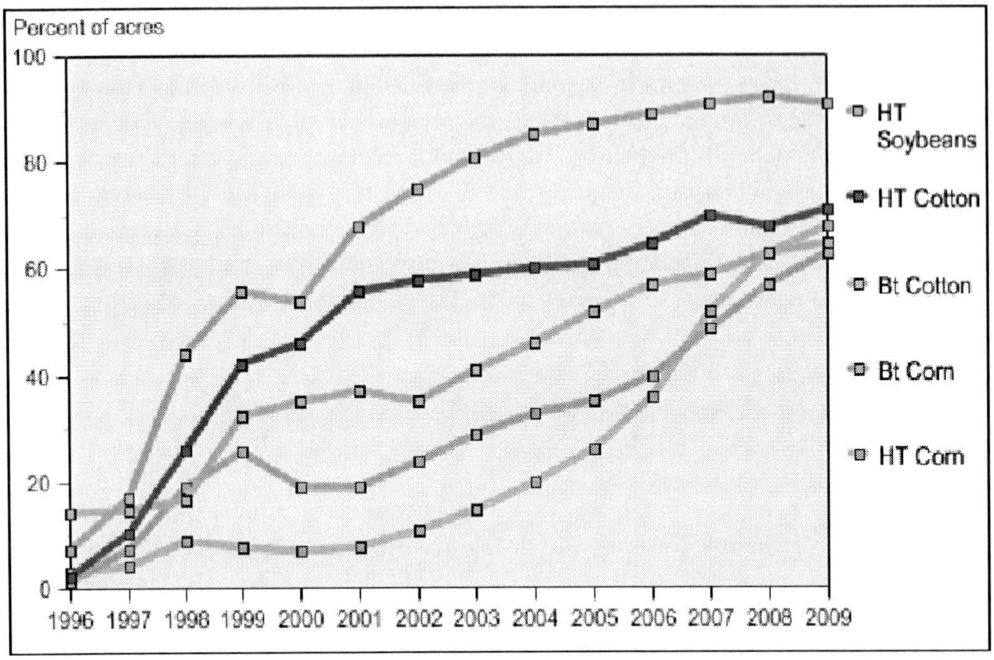

Source: USDA Economic Research Service

Notes: Data for each crop includes more recently developed varieties engineered with both herbicide tolerance and pest resistance. These multiple trait plants are referred to as "stacked trait" varieties.

Figure 2.Adoption of Genetically Engineered Crops in the United States, 1996-2009

Current Applications

Crops

In 2008, GE crops were planted on an estimated 308 8 million acres worldwide, a year-over-year increase of 26.4 million acres (**Figure 1**). The total number of countries growing such crops reached 25 in 2008. Most of the acreage was highly concentrated among four crops—soybeans, corn, cotton, and canola—and six countries. The United States has approximately 50% of global acreage (154.4 million acres), and Argentina had 16.8% (51.9 million acres). Brazil (12.6%, 39.0 million acres), Canada (6.1%, 18.8 million acres), India (6.1%, 18 8 million acres), and China (3.0%, 9.4 million acres) have the largest shares of the remaining planted acres.[4]

In the United States, over 60 GE plant varieties were approved by APHIS for commercial use through early 2005.[5] By 2009, 70 plant varieties had been deregulated by APHIS. Ninety-one percent of all U.S. soybean, 88% of all upland cotton, and 85% of all corn acres were planted with GE seed varieties in 2009, according to USDA's National Agricultural Statistics Service (**Figure 2, Table 1**). Virtually all current commercial applications benefit the production side of agriculture, with weed and insect control by far the most widespread uses of GE crops in the United States and abroad.

Table I. U.S.Acreage in Major GE Crops, 1996 and 2008-2009 (acres in millions)

	Soybeans		Upland Cotton (UC)		Corn	
	Acres	% of all soy planted	Acres	% of all UC planted	Acres	% of all corn planted
1996	4.2	7	2.2	17	2.9	4
2008	68.6	92	7.7	86	69.9	80
2009	77.5	91	7.8	88	73.9	85

Source: USDA-NASS. Acreage Report, June 2009.

Herbicide-tolerant (HT) crops are engineered to tolerate herbicides that would otherwise kill them along with the targeted weeds. These include HT soybeans, HT upland cotton, and to a lesser extent, HT corn. Many of these are referred to as "Roundup Ready" because they are engineered to resist Monsanto's glyphosate herbicide, marketed under the brand name "Roundup." More recently, Monsanto has announced various "stacked trait" varieties—varieties that combine resistance not only to glyphosate/Roundup but also to the herbicides dicamba and glufosinate.

Insect-resistant crops effectively have the pesticide inserted into the plants themselves to control insect pests for the life of the crop. These varieties are often referred to as having a plant- incorporated protectant (PIP). Many of these crops have been genetically engineered with *Bt (Bacillus thuringiensis,* a soil bacterium), which produces a naturally occurring pesticide.[6] These insect-resistant varieties are most prevalent in upland cotton to control tobacco budworm, bollworm, and pink bollworm; and in corn to control earworm and several types of corn borers. Monsanto is also developing "stacked trait" varieties of soybeans and sugar cane that are resistant to insects as well as glyphosate/Roundup.

Other crops approved for commercialization have included varieties of flax, papaya, potatoes, radicchio, rapeseed, rice, squash, sugar beets, tobacco, and tomatoes. However, these are either not commercialized or not widely planted. For example, the biotechnology firm Calgene's FlavrSavr tomato, first marketed to consumers from 1995 to 1997, was withdrawn after Calgene determined that the varieties being grown were not of consistently high quality. GE potato varieties peaked several years ago at 2%-3% of the market; they were discontinued by the seed developer in 2001, mainly after several fast food and snack food companies declined to buy them. Varieties of GE wheat and rice, as well as flax and radicchio, have received regulatory approval but have not been commercially marketed (and/or research has been discontinued), presumably due largely to perceived producer or consumer unease with them.

In contrast to abandoning certain approved GE products, a variety of white GE corn has recently begun to be used in tortilla making after initial resistance by food processors. Herbicide resistant GE sugar beets were only planted in large acreage in the 2008 crop year. While commercially available since 2000, Western beet growers did not plant them because sugar-using food companies (e.g., Hershey, Mars) and beet sugar industry groups (e.g., American Crystal Sugar) balked at the idea of GE beets, thinking that consumers would be opposed. That opposition—real or potential—has apparently subsided to the point that processors have cleared their growers to plant the GE variety.[7] Nonetheless, the Center for Food Safety filed suit in January 2008 challenging APHIS's deregulation of GE sugar beets arguing that wind-pollinated GE sugar beets will inevitably cross-pollinate with related crops being grown in proximity, contaminating conventional sugar beets and organic chard and table beet crops.[8]

Nonetheless, USDA reported that between 1987 and early 2005, APHIS had approved more than 10,700 applications to conduct field tests of various GE crop varieties (out of 11,600 received from companies and other researchers), which the USDA characterized as "a useful indicator of R&D efforts on crop biotechnology." Nearly 5,000 applications were approved for corn alone, followed by soybeans, potatoes, cotton, tomatoes, and wheat. More than 6,700 applications were for HT and insect resistant varieties; the others were to test product quality, virus or fungal resistance, or agronomic (e.g., drought resistance) properties.[9] By October 2008, APHIS had approved more than 13,000 field trials of GE plants, most of which continued to be crop plants bearing genes conferring resistance to certain insects or tolerance to certain herbicides.

Animal Products

Fewer animal-based GE products are commercially available, notably excepting dairy production. Chymosin, a biotechnology-produced enzyme, is used widely in cheese production. Bovine somatotropin (BST, also known as "bovine growth hormone") is a naturally occurring protein that can be produced in greater quantities through genetic engineering. The GE version of BST (rBST) was first approved by the U.S. Food and Drug Administration (FDA) in 1993. Reports suggest that more than 30% of all U.S. dairy cows are administered BST to boost milk production (by an estimated 10%-15%).[10] Several other emerging animal biotechnologies, while not yet commercialized, are believed by researchers to hold great promise (see "Future GE Applications," below).[11] In February 2009, FDA approved the first product from a transgenic animal, an anti- clotting protein derived from the milk of transgenic goats.[12] The animals are genetically engineered to produce a recombinant human antithrombin III protein in their milk.[13]A Netherlands-based biotechnology firm also announced plans to seek U.S. and European approval in 2009 for Rhucin, made from a human protein purified from the milk of genetically engineered rabbits. The protein, Cl esterase inhibitor, helps control inflammation.

U.S. Food Products Containing GE Crops[14]

An estimated 70% of all processed U.S. foods likely contain some GE material. That is largely because two such plants (corn and soybeans, where farmers have widely adopted GE varieties) are used in many different processed foods. U.S. biotechnology rules do not require segregation and labeling of GE crops and foods, as long as they are substantially equivalent to those produced by more conventional methods (see "Regulation and Oversight," below).

Soy-based ingredients include oil, flour, lecithin, and protein extracts. Corn-based ingredients include corn meal and corn syrups, used in many processed products. Canola oil (mostly imported from Canada, where GE-canola is grown) and cottonseed oil are used in cooking oils, salad dressings, snack foods, and other supermarket items. No GE-produced animals are yet approved for human consumption, although cheeses may contain chymosin, and dairy products may have been produced from milk containing GE-BST.

As noted earlier, because most other government-approved GE crops are not being grown commercially, few other GE-derived foods are reaching consumers.

Analysts say some farmers are wary of planting GE crop varieties because their customers may be worried about their safety, although as the case of sugar beets noted above suggests, public opposition to GE products in processed food may be declining. Biotechnology supporters contend that safety concerns are unfounded because scientific reviews have found approved GE crop varieties to be safe, and that foreign governments are simply using such concerns to maintain barriers to imports.

Future GE Applications[15]

"Input" Traits

For farmers, new insect-resistant and herbicide-tolerant GE varieties are under development or have been developed for other crops besides corn, cotton, and soybeans. These include wheat and rice (see below), alfalfa, peanuts, sunflowers, forestry products, sugarcane, apples, bananas, lettuce, strawberries, and eventually other fruits and vegetables. Other traits being developed through genetic engineering include drought and frost tolerance, enhanced photosynthesis, and more efficient use of nitrogen. Tomatoes that can be grown in salty soils, and recreational turf grasses that are herbicide tolerant, pest resistant, and/or more heat and drought tolerant, also are under development. In animal agriculture, pigs have been engineered for increased sow milk output to produce faster-growing piglets. Cloned cattle also have been developed to resist mastitis. APHIS approved field trials in June 2007 for a transgenic sunflower with a carp growth hormone inserted. The GE sunflower would be used in aquaculture feed for farm raised shrimp. Currently awaiting government approval for food use are GE salmon that require as little as half the usual time to grow to market size. Other such fish could follow later.[16]

"Output" Traits

For processors and consumers, research on a range of GE products is continuing: oilseeds low in saturated and transfats; tomatoes with anti-cancer agents; grains with optimal levels of amino acids; rice with elevated iron levels; and rice with beta-carotene, a precursor of Vitamin A ("golden" rice). Other future products could include "low-calorie" sugar beets; strawberries and corn with higher sugar content to improve flavor; colored cotton; improved cotton fiber; delayed- ripening melons, bananas, strawberries, raspberries, and other produce (delayed-ripening tomatoes already are approved); and naturally decaffeinated coffee. Critics, however, point out that, although biotechnology advocates have been forecasting the adoption of various "output" traits for some time, few have actually reached the marketplace.

Other plants being developed could become "factories" for pharmaceutical compounds. The compounds would be extracted and purified for human and animal health uses (among concerns are whether they could "contaminate" food crops; see "Plant-Based Pharmaceuticals from Biotechnology," below). Some varieties of plants under development could also produce "bioindustrials," including plastics and polyurethane. Future transgenic livestock also might yield pharmaceuticals and/or human organ and tissue replacements. To date, none of these innovations have been commercialized, although some have begun field trials.

Regulation and Oversight

Coordinated Framework for Regulation of Biotechnology

The basic federal guidance for regulating biotechnology products is the Coordinated Framework for Regulation of Biotechnology *(51 Fed. Reg.* 23302), published in 1986 by the White House Office of Science and Technology Policy (OSTP). A key regulatory principle is that genetically engineered products should continue to be regulated according to their characteristics and unique features, not their production method—that is, whether or not they were created through biotechnology. The framework provides a regulatory approach intended to ensure the safety of biotechnology research and products, using existing statutory authority and previous agency experience with traditional breeding techniques. The three lead agencies are USDA's Animal and Plant Health Inspection Service (APHIS), the Food and Drug Administration (FDA) at the Department of Health and Human Services, and the Environmental Protection Agency (EPA).

Animal and Plant Health Inspection Service

APHIS regulates the importation, interstate movement, and field testing of GE plants and organisms that are or might be plant pests under the Plant Protection Act (PPA; 7 U.S.C. §7701 *et seq.).* APHIS regulates animal biologics (i.e., viruses, serums, toxins for animal vaccines) under the Virus, Serum, and Toxins Act (21 U.S.C. 151 *et seq.).* Specifically, GE plants that are or might be plant pests are considered "regulated articles" under APHIS regulations (7 CFR 340-340.9). APHIS authorization must be obtained prior to import, interstate movement, or environmental release, including field testing.

More specifically, a "regulated" plant cannot be introduced into the environment, or even field tested, unless its developer obtains APHIS authorization through either the (1) permit process or (2) notification process. Permits impose restrictions on movement and planting to prevent escape of plant material that may pose a pest risk. Sponsors follow APHIS guidance on testing and movements to ensure that the plant will not damage agriculture, human health, or the environment. Plant-based pharmaceuticals virtually always must be developed under the permit process. However, most other GE crops have been developed under the notification option, an expedited procedure that is less rigorous than permitting. Notification can be used in lieu of permitting when the plant species is not considered a noxious weed (or weed in the release area) and other APHIS standards are met.

Regardless of the process chosen, after testing is completed, a developer next seeks "non-regulated status" from APHIS, the typical route to full commercialization and no further formal oversight. The developer must provide APHIS with extensive information on plant biology and genetics, and potential environmental and plant pest impacts that may result from the modification. APHIS conducts a formal environmental assessment (EA) and has public comment periods before deciding whether to approve the developer's request for "non-regulated status."

Food and Drug Administration (FDA)

FDA regulates food, animal feed additives, and human and animal drugs, including those from biotechnology, primarily to ensure that they pose no human health risks, mainly under the Federal Food, Drug and Cosmetic Act (FFDCA; 21 U.S.C. §301 *et seq.)* and the Public Health Service Act (42 U.S.C. §201 *et seq.)*. Under the FFDCA, all food and feed manufacturers must ensure that the domestic and imported products they market are safe and properly labeled. All domestic and imported foods and feeds, whether or not they are derived from GE crops, must meet the same standards. Any food additive, including any introduced through biotechnology, cannot be marketed before it receives FDA approval. However, additives that have been determined to be "generally recognized as safe" (GRAS) do not need such preapproval.

To help sponsors of foods and feeds derived from GE crops comply, FDA encourages them to participate in its voluntary consultation process. All GE-derived products now on the U.S. market have undergone this process. With one exception, none of these foods and feeds were considered to contain a food additive, so they did not require approval prior to marketing. However, a May 1992 FDA policy statement noted that GE foods must undergo a special review under certain conditions, such as if the gene transfer produces unexpected genetic effects, changes nutrients or toxicant levels from the food's traditional variety, might contain an allergen from another crop, or would be used to host an industrial or pharmaceutical substance, for example.[17]

In June 2006, FDA published new guidance under which developers of new plant varieties intended for food use—including those that are bioengineered—can provide FDA with any information about new proteins they are using in the early stages of crop development. This voluntary consultation is to occur prior to the stage of development where the new proteins might "inadvertently" enter the food supply. FDA believes that any potential risk from the low-level presence of such material in the food supply would be limited to the remote possibility of it containing or consisting of a new protein that might be an allergen or toxin.[18]

On January 15, 2009, the U.S. Food and Drug Administration (FDA) released final guidance on how it is to regulate GE animals and products. FDA is to do so under its existing statutory authority and regulations. Generally, GE-derived foods, for example, are to be regulated like non- GE foods; if their composition does not differ from their conventional counterparts, they will not have to be labeled. Nonetheless, developers of GE animals and of GE-derived products must gain FDA pre-market approval.

Although animal biotechnology involves many techniques other than cloning, this latter technology has attracted widespread attention. A final risk assessment and industry guidance on the safety of meat and milk from cloned cattle, pigs, and goats and their offspring were released January 15, 2008, by FDA. The documents generally echoed FDA's December 28, 2006, draft risk assessment, which found that such products are as safe to eat as those of conventionally bred animals. The FDA also concluded that cloning poses the same risks to animal health as those found in animals created through other assisted reproductive technologies—although the frequency of such problems is higher in cloning. (Scientists stress that cloning is an assisted reproduction technique that does not involve any transfer or alteration of genes through GE.) The agency said it was no longer asking industry to refrain voluntarily from marketing the products of cloned animals and their offspring, although the U.S. Department of Agriculture (USDA) did ask that it be continued for products from clones (but not from the offspring of clones).[19]

Environmental Protection Agency (EPA)

EPA must approve the use of all pesticides, including those genetically engineered into plants, which it terms "plant-incorporated protectants" (PIPs). EPA essentially determines a PIP's environmental safety through its authority under the Federal Insecticide, Fungicide, and Rodenticide Act (FIFRA; 7 U.S.C. §136 *et seq.).* Also, under the FFDCA, the EPA establishes tolerances (i.e., safe levels) for pesticides in foods. Pre-commercial regulation is through a system of notifications for small-scale field tests or experimental use permits for larger field tests. As for any pesticide, EPA requires the manufacturer of a PIP to obtain a registration through a regulatory process intended to ensure its safe use environmentally.

In practice, all three agencies have more detailed procedures than described here for monitoring and approving the development and commercialization of GE crops and foods, particularly if they are for new uses (e.g., pharmaceuticals). However, the fundamental guiding policy assumption since 1986 has been that the biotechnology process poses no unique or special risks; therefore it demands no new laws beyond those that already govern the health, safety, efficacy, and environmental impacts of more traditional production methods.

Assessments of Current Policy

The biotechnology industry, prominent U.S. agricultural groups, and many scientific authorities continue to subscribe to the current coordinated framework described above. They cite various studies in asserting that there is no evidence that current GE crops have harmed the environment or human health.[20]

These reports generally conclude that current GE crops likely pose no greater risks than conventional varieties, that each GE product should be assessed on a case-by-case basis, and that the current U.S. regulatory framework is adequate. However, the reports have also suggested a number of administrative or regulatory changes that might be adopted to improve oversight.

Critics, including some consumer and environmental groups, have gone further, raising questions about whether the current laws themselves remain adequate to protect human health and the environment, particularly as emerging GE applications—such as plant-based pharmaceuticals and industrial compounds, and transgenic animals, including insects—increasingly challenge the agencies' regulatory capabilities. They see gaps in the existing pre-market approval processes, and in post-market oversight of GE crops, that they contend may expose humans and the environment to unwarranted risks These critics have argued that new legislation is needed to clarify agency roles and strengthen their regulatory authority, particularly over future novel GE applications.

A number of agricultural organizations, while not necessarily clamoring for new laws, have expressed wariness about some new biotechnology products now awaiting approval. Among other concerns, they worry about consumer acceptance, potential difficulties exporting these varieties to countries demanding the segregation and labeling of GMOs (or outright prohibition of GMOs), and the potential for inadvertently mixing GE with non-GE crops. The 2006 discovery of an unapproved variety of GE rice in commercial U.S. rice supplies, and the 2008 discovery of an unapproved GE cotton variety harvested with an approved variety, are indicative of the problem.

Another court case involving the deregulation of Monsanto's GE alfalfa also raised important concerns about the adequacy of APHIS regulatory regime. In May 2007, U.S. District Court for the Northern District of California in San Francisco held that APHIS had failed to properly consider the environmental effects of the GE alfalfa in granting approval.[21] A coalition of farmers, consumers, and environmentalists, led by the Center for Food Safety, filed suit in 2006 alleging that GE alfalfa could create "super weeds" resistant to herbicide, hurt production of organic dairy and beef products, and cause farmers to lose export business due to risks of contamination to natural and organic alfalfa. Perhaps more than some other GE varieties, the GE alfalfa case raised important issues about the limitations of coexistence between traditional and GE production methods. The issue of whether gene flow from GE alfalfa could permanently harm growers who did not want to adopt GE varieties case was particularly clear in this case.[22]

USDA Advisory Committee Report

In late August 2006, USDA released a long-awaited status report by its Advisory Committee on Biotechnology and 21st Century Agriculture (AC21). The report covered biotech adoption and regulation, and included a discussion of the many outstanding policy issues. The AC21 report observed, for example, that "U.S. regulations are evolving slowly and many governing statutes were written before modern agricultural biotechnology was developed. That system may not be optimal to meet the needs of producers and consumers."[23]

Although all the AC21 members agreed on the importance of ensuring the food and feed safety of transgenic crops, they had differing views "about whether the current FDA regulatory system for transgenic crops was adequate to ensure safety and public acceptance." Among other observations, the AC21 cited the lack of a "clear, comprehensive federal regulatory system to assess the environmental and food safety of transgenic animals before they are commercialized."

All sides of the debate, however, continue to agree that whatever policy course is pursued in the future, it should provide for a clear, predictable, trusted regulatory process.[24]

Views on the FDA Guidance

The recent FDA guidance on early food safety evaluations for new plant varieties (issued in June 2006; see page 8) is widely viewed as that agency's current policy thinking on AP. The Biotechnology Industry Organization (BIO) supported the FDA guidance, noting that it "provides safety assurance, while also recognizing the fact that 'adventitious presence' is a natural part of plant biology, seed production, and the distribution of commodity crops." Several food industry officials also characterized the guidance as an important step toward a science-based policy regarding AP. However, critics such as the Center for Food Safety (CFS), a food safety and environmental advocacy organization, have complained that the guidance will more likely encourage "contamination" of the food supply by GE varieties rather than improve safety oversight. Moreover, the policy does not attempt to define or quantify an acceptable level, or levels, of AP.[25]

In 2006, CFS sued FDA for allegedly failing to adopt any pre-market safety requirements for GE foods, or to require labels identifying foods containing GE material. The lawsuit sought the establishment of a mandatory, pre-market review system for all such foods.[26] The case was subsequently dropped by agreement with the parties.

APHIS Oversight

USDA's APHIS has taken a number of actions over the past several years intended to improve regulatory oversight (like FDA, utilizing its current legislative authorities). These have included consolidation of its activities under a new Biotechnology Regulatory Services (BRS) office; development of a compliance and enforcement unit to ensure GE developers' adherence to the rules, and the publication of more stringent permit conditions for GE-derived plants for pharmaceuticals and industrials (see "Plant-Based Pharmaceuticals from Biotechnology," below).

In the January 23, 2004, *Federal Register,* the agency published a notice of its intent to prepare an environmental impact statement (EIS) evaluating these regulations, and requesting public comment on a number of possible changes. These include whether to broaden APHIS's regulatory scope to cover GE plants that may pose a noxious weed risk or may be used as biological control agents; whether to establish new categories for field testing that delineate requirements based upon relative levels of potential risk; and whether to change (i.e., strengthen) its environmental reviews and permit conditions for GE plants producing pharmaceuticals and industrials. APHIS also solicited comments on ways that it might ease its requirements for lower-risk products. The agency received over 3,000 comments on its proposal.

OIG Criticisms

In a December 2005 audit report, USDA's Office of Inspector General (OIG) criticized APHIS's biotech regulation. Noting the approval, at that point, of more than 10,600 applications for GE tests at more than 49,300 field sites, the OIG expressed concern that "the Department's efforts to regulate those crops have not kept pace." Various weaknesses in the approval and inspection process "increase the risk that regulated genetically engineered organisms will inadvertently persist in the environment before they are deemed safe to grow without regulation," the report observed.[27]

More specifically, the OIG stated that APHIS lacked basic information about the field test sites that it has approved, including their precise locations; and about what becomes of the crops— including those tested for pharmaceutical or industrial uses—after testing ends. Where notifications (rather than permits) were used, APHIS did not review applicants' containment protocols. Among other things, the OIG noted that APHIS site inspection requirements were vague and not always fulfilled by inspectors, and that the agency's guidance for containing GE crops and seeds needed strengthening.

Responding to the audit report, APHIS stated that most of the OIG recommendations "reaffirm APHIS' decision to create the new Biotechnology Regulatory Service (BRS) and devote greater resources toward regulating biotechnology. Most of the recommendations are in line with changes that BRS has already enforced, is currently undertaking, or plans to implement."[28]

In January 2009, the OIG released another report concluding that the department did not have an import control policy to regulate GE animals and that its import policy for GE crops could become outdated as other countries increase the number of biotechnology products.[29]

APHIS Regulatory Changes

In July 2007, APHIS published a draft environmental impact statement (EIS) as part of the evaluation of its regulatory structure. In October 2008, APHIS proposed a revision of its regulations regarding the importation, interstate movement, and environmental release of certain GE organisms. [30] The public comment period initially was to end on November 24, 2008, but was extended to June 29, 2009. A subsequent issue-focused meeting on the proposed rule changes was held on April 29-30, 2009. The final rule has not yet been published. These proposed revisions are the first since the regulations were established in 1987. Under current regulations, a GE organism is a regulated article if it is a plant pest or there is reason to believe it might become a plant pest. In the notification of the proposed regulation revisions, APHIS stated that technological advances have led to the possibility of developing GE organisms that do not fit within the plant pest definition, but still might cause environmental or other physical harm by the definition of a plant pest under the Plant Protection Act. According to APHIS, the new regulations would subject a GE organism to oversight based upon known plant pest and noxious weed risks of the parent organisms, or based upon the traits of the GE organism, or based upon the possibility of unknown risks as a plant pest or noxious weed when insufficient information is available.[31] The proposed regulations also include regulating GE seedlings, tubers, cuttings, bulbs, spores, etc.

APHIS further proposes to reorganize the regulations for permit applications and evaluation procedures by discontinuing its notification procedure, while retaining the permitting procedure. The proposed regulations would also establish a new petition procedure for APHIS to approve a new conditional exemption from the permit requirements, which is currently done by amending regulations.

For environmental releases, APHIS proposes a permitting system based on two primary risk- related factors: (1) the ability of the unmodified recipient plant species to persist in the wild and (2) the potential of the GE trait to cause harm based on the plant pest and noxious weed definitions. With respect to the persistence factor, APHIS proposes grouping plant species into four risk categories based on the risk of persistence of the plant or its progeny in the environment without human intervention. Four similar risk categories are also proposed for potential harm caused by the GE trait. Other proposed regulatory changes include remediation authorities for failure to comply with regulations and agency response to low-level presence (LLP) of regulated plant materials in commercial seeds or grain that may be used for food or feed.

Reactions to the proposed revisions were mixed, and were, in part, the reason APHIS extended the original comment period and held public meetings on some of the more controversial proposed changes (e.g., scope of the regulatory changes, incorporation of the Plant Protection Act's noxious weed authority into APHIS 's regulatory authority, revision of the permit process, and environmental release of GE crops that produce pharmaceutical and industrial compounds). In their comments on the proposed rule changes, biotechnology industry representatives and nongovernmental organizations expressed opposition to the expansion of APHIS authority to regulate GE organisms if they posed a risk as a plant pest or noxious weed. The industry representatives also took issue with the with the proposal to take a voluntary approach to GE regulation, arguing that it could have a significant impact on international trade.[32] The Center for Food Safety (CFS) denounced the proposal, stating that "these proposed regulations may set in motion a process that would put many GE crops completely beyond the bounds of regulation." CFS said that its biggest concern is that the proposed rules remove established criteria in determining the very scope of regulation. In a similar response, the Union of Concerned Scientists denounced the proposed rules for failing to adequately protect the U.S. food supply from potential contamination from biopharm crops through cross-pollination or seed mixing between biopharm food crops and those food crops intended for consumption.[33] In March 2009, more than 80 advocacy groups signed a letter urging Secretary of Agriculture Tom Vilsack to halt approving GE crops until the agency changes its regulatory approach to biotechnology.

The National Environmental Protection Act and APHIS

After a GE variety is approved for release into the environment on a test basis, the owner of the GE seed generally petitions APHIS for "deregulated status" of the particular GE "event" that has been approved. This is the last step to full-scale commercialization of the GE plant. Once the GE plant is deregulated, it is no longer subject to APHIS regulation under 7 C.F.R. Part 340. A significant step in the deregulation process involves an assessment of the plant's environmental impact. The National Environmental Policy Act (NEPA) requires federal agencies to prepare a detailed Environmental Impact Statement (EIS) for all "major Federal actions significantly affecting the quality of the human environment." The regulations governing the fording of a significant effect are promulgated by the Council on Environmental Quality (CEQ). When an EIS is not categorically required, an agency may determine, from the data already at hand, that the environmental impacts are not significant enough to warrant an EIS. This judgment permits the agency initially to prepare an Environmental Assessment (EA) rather than the lengthier and more detailed EIS. An EA is a public document that briefly provides the basis for determining whether to move forward with an EIS or to make a finding of no significant impact (FONSI).

Several cases over the past five years have raised issues about the adequacy of APHIS's regulatory structure in moving to deregulate a GE plant. APHIS on several occasions has issued a FONSI on the basis of an EA and deregulated a GE plant, only to have that decision decisively challenged in court.

GE Alfalfa

A U.S. District Court held in February 2007 that APHIS failed to properly consider the environmental effects of Monsanto's GE alfalfa. The court vacated APHIS's June 2005 decision deregulating GE alfalfa on the basis of an EA. In March 2007, the District Court of the Northern District of California issued a preliminary injunction, and in May 2007 the court issued a permanent injunction against planting or selling Monsanto's line of GE alfalfa until a final EIS was prepared. In June 2009, the Ninth Circuit Court of Appeals affirmed the illegality of APHIS's approval of deregulated status for Monsanto's GE alfalfa.

Not only did the GE alfalfa have environmental implications for domestic producers, the suit, brought by a coalition of farmers and the Center for Food Safety, also cited the concerns of farmers who sell to export markets.[34] Japan and South Korea, America's most important alfalfa customers, have warned that they will discontinue imports of U.S. alfalfa if a GE variety is grown in this country. U.S. alfalfa exports total nearly $480 million per year, with about 75% going to Japan. The court disagreed with USDA's assertion that exports to Japan would not be harmed by the deregulation of GE alfalfa.

On January 12, 2010, APHIS announced that the draft environmental impact statement concerning Monsanto and Forage Genetics International lines of GE alfalfa was available for public comment.[35] A series of public meetings in several cities was also held on February 3, 4, and 9, 2010. After a three-week extension for comments, the comment period ended March 3, 2010. The final EIS, which will address the nearly 135,000 comments received, is expected to be published before the end of the year. APHIS has stated that the EIS will be completed in time for spring 2011 planting.

The court's decision to enjoin planting GE alfalfa until the final EIS was published was appealed to the Supreme Court. [36] The Court agreed to hear that case on January 15, 2010, and on April 27, 2010, the Court heard oral arguments. Monsanto's appeal concerned whether the federal courts in the Ninth Circuit properly applied and interpreted the injunction standard in NEPA cases when they permanently enjoined planting until the EIS was complete. On June 21, 2010, in a 7-1 opinion written by Justice Alito, the Supreme Court reversed the injunction, saying that the Ninth Court had overreached itself procedurally in halting the plantings. As of late August 2010, APHIS has not announced any intention to re-authorize the restricted plantings. Were APHIS to do so, any planting of GE alfalfa would have to be done under a rule for "partial deregulation," a modification of existing planting restrictions for regulated organisms, perhaps limiting planting to specific geographic areas under controlled procedures. Such a partial deregulation, however, could trigger the need for a new EA on the impact of partial deregulation.

GE Sugar Beets

In February 2005, APHIS issued a fmding of no significant impact on the environment (FONSI) for the cultivation and agricultural use of a Monsanto-developed variety of glyphosate-tolerant sugar beet (Event H7-1).[37] The ruling meant that GE beets were no longer a regulated article under 7 C.F.R. Part 340. The GE beets were first planted in the western United States in spring 2008. Similar to the response to APHIS's FONSI regarding GE alfalfa, a case was filed in U.S. District Court for the Northern District of California by the Center for Food Safety and Earthjustice in January 2008 representing a coalition of farmers and consumers.[38] As with the alfalfa case, the plaintiffs claimed that APHIS had violated NEPA by not conducting an EIS before granting the GE beets unregulated status.

Sugar beet seed is grown primarily in Oregon's Willamette Valley. The area is also an important seed growing area for crops closely related to sugar beet (e.g., chard and table beets). Sugar beets, like alfalfa, are wind pollinated. Eventually, GE beets would cross-pollinate with related crops. This would have significant economic impacts on organic producers of Swiss chard and table beets being grown in the same areas as the GE beets. In his September 2009 order requiring APHIS to prepare an EIS, the presiding judge determined that "the potential elimination of a farmer's choice to grow non-genetically engineered crops, or a consumer's choice to eat non- genetically engineered food, [is] an action that potentially eliminates or reduces the availability of a particular plant [and] has a significant effect on the human environment." The court concluded that there was no support in the record for APHIS's conclusion of no significant impact.

At a meeting in December 2009, the U.S. District Court for the Northern District of California set out a schedule and process for determining the remedies regarding the sugar beets. The process did not interfere with the planting of the beets in 2010. Oral arguments were heard on June 11, 2010. On August 13, 2010, the court revisited the issue of whether to issue an injunction, as had been done with GE alfalfa. Given that the Supreme Court had just ruled that the injunction for GE alfalfa was improper, the Ninth Court vacated APHIS's deregulation decision without enjoining planting. While avoiding the "drastic remedy," as the Supreme Court described the injunction of GE alfalfa, the practical effect of vacating APHIS's GE sugar beet deregulation decision is that planting GE sugar beet is now effectively halted. Because nearly 95% of sugar beets planted in 2009/2010 are the GE variety, this raises questions about the availability of non-GE sugar beet seed for the 2011 planting season.

GE Ethanol Corn

In June 2009, APHIS filed a request for additional comments on a petition by Syngenta Seeds, Inc. to deregulate Alpha-Amylase Maize Event 3272.[39] This GE variety of corn produces a microbial enzyme that facilitates ethanol production. APHIS has prepared a draft EA and plant pest risk assessment for review and comment. Most relevant in this request is that it is the first request to deregulate a GE plant variety that is intended solely for the use in the production of ethanol. The corn variety is not cultivated for human consumption or livestock feed, but rather is grown to improve the efficiency of converting corn starch to industrial ethanol. A concern is that there is inadequate scientific data or documentation to evaluate the possible impacts on food and feed should this variety be commingled with commodity corn supplies.[40]

GE Eucalyptus

APHIS granted a permit on May 12, 2010, to import a GE eucalyptus tree that is "cold-tolerant." If commercialized, the hybrid Eucalyptus would be used for pulp and biofuel production. Eucalyptus is a fast-growing tree that dominates tropical timber plantations. It is not native to the United States and has become invasive in some places. The permit was issued to a company called ArborGen, LLC. ArborGen is a joint initiative of International Paper, MeadWestvaco, and Rubicon. The permit authorizes planting and flowering on 28 sites across seven southern states (Alabama, Florida, Georgia, Louisiana, Mississippi, South Carolina, and Texas). The Center for Biological Diversity and other organizations have sued to set aside the approval on grounds that the potential environmental impacts have not been properly evaluated, nor has APHIS complied with congressional mandates enacted in the 2008 farm bill (P.L. 110-246) requiring more rigorous oversight of field testing for GE organisms.[41]

GLOBAL TRADE CONCERNS

The U.S. approach to biotechnology regulation contrasts with that of many major trading partners. For example, the European Union (EU), Japan, South Korea, New Zealand, and Australia either have or are establishing separate mandatory labeling requirements for products containing genetically modified ingredients; in many of these countries, consumer and official attitudes toward GE foods are more skeptical. Differing regulatory approaches have arisen at least partly because widely accepted international standards continue to evolve. Incidents, such as those discussed below, have disrupted U.S. exports and contributed to trade tensions.[42]

GE Rice

Although several GE varieties of rice have been approved for commercial use ("deregulated," in regulatory parlance), none have been marketed, although they have been planted on test plots in the United States. In August 2006, the Secretary of Agriculture announced that "trace amounts" of an unapproved variety of GE rice had been found in samples of the 2005 crop of U.S. long grain rice. The Secretary and other USDA officials sought to reassure the rice trade and consumers that the findings posed no human health, food safety, or environmental concerns.

Owner Bayer CropScience had not asked APHIS to deregulate this particular line, called LLRICE601, which had been field tested between 1998 and 2001. Two other Bayer GE rice varieties, known as LLRICE62 and LLRICE06, had received commercial approval but have not been commercialized, USDA stated. Also, "[t]he protein in LLRICE601 is approved for use in other products" and "has been repeatedly and thoroughly scientifically reviewed and used safely in food and feed, cultivation, import and breeding in the United States, as well as nearly a dozen other countries around the world."[43]

Nonetheless, the discovery unsettled rice markets and rekindled longtime criticisms of U.S. biotechnology regulatory policies. The U.S. rice crop is valued at nearly $2 billion annually. Exports represent approximately one-half or more of U.S. rice production annually on a volume basis, of which about 80% is long grain (the type in which GE material was detected), according to USDA statistics. Although the United States produces only about 1.5%-2% of the world rice crop, it was the fourth-leading exporter (behind Thailand, Vietnam, and India), with more than 13% of world market share in 2005.

Of the 4 4 million metric tons (MMT) exported in 2005, Mexico was by far the leading buyer, at 753,000 MT. Japan was the second-leading market at nearly 424,000 MT. Various Central American and Caribbean countries took a total of 1.4 MMT; Iraq, 310,000 MT; and European Union (EU) countries, a total of 306,000 MT, USDA data show. Much of the long grain crop is produced in southern U.S. states, which generally ship from Gulf ports to Latin America, the Caribbean, and Europe, for example. California grows mainly medium and short grain rice varieties, which are marketed in Asia, including Japan.

Following USDA's notification that U.S. rice supplies had traces of GE material, September 2006 closing rice futures dropped from $9.70 per cwt. (100 pounds) on August 18, closing at $8.99 per cwt. on August 25, 2005. (One year ago, the closing price was less than

$7.00 per cwt.) The European Union (EU), which bought 279,300 MT of U.S. long grain rice in 2005, reacted by adopting a measure requiring all such shipments to be tested and certified as free of LLRICE601. Japan has indicated that it was suspending shipments of U.S. long grain rice although, as noted, most U.S. rice exports there are short and medium grain.

According to a statement by the producer cooperative Riceland Foods, Inc., of Stuttgart, AR, the GE material was initially discovered by one of its export customers in January 2006. Riceland then sent a sample to a U.S. laboratory, which confirmed the Bayer GE trait, which is known to be present in (and approved for) corn, soybeans, canola, and cotton. Riceland said it collected samples from several storage locations in May 2006 and found positive results that were "geographically dispersed and random throughout the rice-growing area." Bayer was notified in early June, and its tests confirmed the presence of the GE trait in the equivalent of 6 per 10,000 kernels (0.06%).[44]

In August 2006, USDA officials offered few additional details about the cause or extent of the problem. They indicated that they had not been informed by Bayer of the discovery until July 31, after which the Department began its own investigation, they stated. Among other actions, USDA said that APHIS was now moving to approve (i.e., deregulate) LLRICE601. Also, USDA's Grain Inspection, Packers, and Stockyards Administration (GIPSA) has verified the use of two standardized tests that can test for the GE protein in rice shipments.

Consumer and environmental advocacy groups were harshly critical of APHIS and USDA, noting that officials waited three weeks to make the discovery public—and still did not know where the samples were grown or how they entered the food supply. One group, the Center for Food Safety, subsequently called for a moratorium on all new field testing permits until oversight can be improved.[45] In August 2006, rice farmers in Arkansas, Missouri, Mississippi, Louisiana, Texas and California filed a class action lawsuit against Bayer CropScience, accusing the company of negligence in allowing unapproved genetically engineered rice to find its way into the commercial supply chain. By November 2006, APHIS declared the rice variety LLRICE601 safe for human consumption and deregulated the variety. USDA essentially declared that the new variety was similar to two Bayer varieties that had already been approved.

GE Wheat

Trade concerns were apparent in the debate over whether to introduce (commercialize) a variety of glyphosate/Roundup-resistant wheat. Monsanto had asked the U.S. and Canadian governments for their approval, and other GE wheat varieties had been under development. Some producers wanted to plant the wheat as soon as it became available; others feared rejection by foreign customers of not only GE wheat, but all U.S. and Canadian wheat, out of concern that even non- GE shipments might unintentionally contain some GE grain. The latter group wanted developers and regulators to wait for more market acceptance before releasing GE wheat varieties.

In early 2003, a group of U.S. wheat producers petitioned the Administration to conduct a more thorough assessment of the environmental impacts of the Monsanto request; 27 farm, religious, and consumer advocacy organizations endorsed the petition in early 2004.

Underlining these concerns, Japanese consumer groups in March 2004 reportedly told U.S. officials in wheat- dependent North Dakota that their country would not import any U.S. wheat products if the Monsanto application was approved.[46]

This resistance likely contributed to a decision by Monsanto to discontinue its efforts to win regulatory approval of a genetically modified wheat variety. Monsanto announced its decision in May 2004. Although Monsanto withdrew its applications for regulatory approval from EPA and APHIS, it did not withdraw its FDA application. FDA subsequently approved the application in July 2004. However, FDA approval alone is not sufficient to bring the GE wheat to market.

While opposition to GE wheat remains strong among many U.S. trading partners, a spokesman for the joint biotechnology committee of the National Association of Wheat Growers and U.S. Wheat Associates, indicated in 2007 that support for planting and exporting GE wheat was growing among U.S some wheat producers.[47]

In May 2009, wheat grower associations in the United States, Canada, and Australia issued a joint statement announcing that they intend to "work toward the goal of synchronized commercialization of biotech traits in our wheat crops.... [W]e believe it is in all of our best interests to introduce biotech wheat varieties in a coordinated fashion."[48] This joint statement produced an immediate reaction by various environmental and consumer organizations. Canada's Farmers Union, the Organic Federation of Australia, the U.S. Organic Consumers Association, and other organizations drafted a statement opposing commercializing GE.[49] In July, Monsanto Canada's spokesperson stated that the future "GE wheat will not be Roundup ready.... It will have increased drought tolerance, increased yield, and improved nitrogen efficiency."[50] Monsanto anticipates that it will be at least 10 years before their research and development efforts produce a commercial variety of GE wheat.

U.S.-EU Dispute

In May 2003, the United States, Canada, and Argentina initiated a complaint before the World Trade Organization (WTO) regarding the EU's de facto moratorium on approvals of new GE crops. U.S. agricultural interests contended that the moratorium not only blocked exports such as corn and other products to the EU, but also was fueling unwarranted concerns about the safety of agricultural biotechnology throughout the world. The United States and its allies further argued that the EU moratorium was violating WTO rules stating that a country's actions to protect health and the environment must be scientifically based, and approval procedures must be operated without undue delay.

The WTO named a panel in March 2004 to consider the case. Although the EU effectively lifted the moratorium in May 2004 by approving a genetically engineered corn variety, the three complainants pursued the case, in part because a number of EU member states have continued to block approved biotech products. In February 2006, the WTO dispute panel, in its interim confidential report, ruled that a moratorium existed, that bans on EU-approved GE crops in six EU member countries (Austria, France, Germany, Greece, Italy, and Luxembourg) violated WTO rules, and that the EU failed to ensure that its approval procedures were conducted without "undue delay." The final ruling was circulated to the parties in May 2006 and made public in September 2006.

The dispute panel's ruling dismissed several other U.S. and co-complainant claims, and did not address such sensitive issues as whether GE products are safe or whether an EU moratorium on GE approvals continued to exist. The final ruling, among other things, directed the EU to bring its practices in line with WTO rules. It concluded that the EU had breached its commitments with respect to 21 products, including types of oilseed rape, maize and cotton. It also said individual bans in Austria, France, Germany, Greece, Italy and Luxembourg were illegal.

The EU initially agreed on a November 2007 deadline for compliance with the WTO dispute ruling. The parties subsequently agreed to extend the time for EU compliance with the ruling to January 2008. The EU missed this deadline in large measure. Brussels has found it hard to implement the WTO ruling because some of the 27 EU member states operate their own bans on GE crops. Individual countries (e.g., Austria, France, Greece) have prohibited the sale or cultivation of certain EU-approved varieties of GE corn (e.g., MON810, a variety produced by Monsanto). In 2008, France also initiated a temporary national moratorium on GE crops. Spain continues to dominate the EU in GE crop cultivation.

Although positive action has been slow, the United States has temporarily suspended WTO sanctions. U.S. agricultural interests, however, remain concerned that the stricter EU rules for labeling and tracing GE products will continue to discriminate against U.S. exports. If progress is not made, the issue is likely to return to the WTO's dispute settlement body. The United States could retaliate against the EU to compensate for the annual value of lost U.S. exports, royalties and licensing fees to the EU from biotech crops. These could be levied by imposing extra tariffs on EU goods or lifting other WTO agreements regulating agriculture or health and safety.

The WTO case did not involve the EU's new "labeling and traceability" regulations, in effect as of April 2004, to require most food, feed, and processed products from GMOs to be labeled. GE- based products also must be segregated from non-GE products, with documentation. U.S. agricultural interests argue that, even if the EU regularly approves GMOs, the labeling and traceability rules are themselves unworkable and unnecessary, and can mislead consumers by wrongly implying that GM-derived products are inherently different than non-GM foods or pose safety concerns.[51] The EU, however, continues to defend its mandatory labeling regime.

At least one EU country, Germany, has addressed the issue of potential liability from GM crops— passing a law in November 2004 that holds farmers who plant GM crops liable for damages to nearby non-GM fields (even if the GM farmers adhered to planting instructions and regulations). Some U.S. interests countered that the moratorium will not effectively end until the EU clears more of some two dozen or more GE food and agricultural products still awaiting regulatory approval—and EU member states actually implement the approvals.

The Biosafety Protocol

The Cartagena Biosafety Protocol, an outgrowth of the 1992 Convention on Biological Diversity (CBD), was adopted in January 2000 and took effect in 2003. The United States is not a party to the 1992 CBD, and therefore cannot be a party to the protocol. However, because its shipments to ratifying countries are affected, it has actively participated in the negotiations over the protocol text and in countries' preparations for implementation.

The protocol, which 134 other nations had ratified as of August 2006, permits a country to require formal prior notifications from countries exporting biotech seeds and living modified organisms (LMOs) intended for introduction into the environment. The protocol requires that shipments of products that may contain LMOs, such as bulk grains, be appropriately labeled and documented, and provides for an international clearinghouse for the exchange of LMO information, among other provisions. The Protocol further establishes a process for considering more detailed identification and documentation of LMO commodities in international trade.

The United States objected to implementing measures approved during an international conference in Kuala Lumpur in February 2004. According to the United States, the measures would mandate overly detailed documentation requirements and potentially expose exporters to unwarranted liability damages if imported GMOs harm the environment or human health. U.S. government and industry officials believe that these and other rules could disrupt U.S. exports.[52]

GMOs in the Developing World

In Asia, particularly China and India, governments view GMOs as a way to produce more food for burgeoning populations, despite some in-country opposition and support for labeling GE products. China has been researching GE corn, cotton, wheat, soy, tomatoes, and peppers since 1986. It has, however, been reluctant to approve commercial varieties of GE, which have been under development there. If so, it would be the first time a GE plant was used widely as a staple food, and may influence the decisions of other Asian countries with regard to accepting GE foods.[53]

In the debate over the potential contribution of biotechnology to food security in developing countries, critics argue that the benefits of biotechnology in such countries have not been established and that the technology poses unacceptable risks They also suggest that intellectual property rights (IPR) protection gives multinational companies control over developing country farmers. Proponents say that the development of GE technology appears to hold great promise, with the potential to complement other, more traditional research methods, as the new driving force for sustained agricultural productivity in the 21st century. They maintain that IPR difficulties have been exaggerated.[54]

According to a recent report published by the International Service for the Acquisition of Agri- biotech Applications, 12 developing nations planted GE crops in 2007. Of the total 114 3 million hectares of GE crops cultivated worldwide, 43% of the global GE crop area is in developing countries.

Differences on this issue were featured in 2002, when the United Nations (UN) World Food Program (WFP) announced an appeal for food aid to meet the needs of some 14 million food- short people in six southern African countries: Lesotho, Malawi, Mozambique, Swaziland, Zambia, and Zimbabwe. However, a debate over the presence of genetically modified corn in U.S. food aid shipments made the provision of food aid more difficult and costly. Some of the countries expressed reluctance to accept unmilled GE corn on account of perceived environmental and commercial risks associated with potential introduction of GE seeds into southern African agriculture. Zambia refused all shipments of food aid with GE corn out of health concerns as well. In March 2004, Angola said it too would ban imports of GE food aid, including thousands of tons of U.S. corn, despite a need to feed approximately 2 million Angolans.

The United States has blamed EU policies for southern African countries' views on food aid containing GE products. The United States maintains that genetically modified crops are safe to eat and that there is little likelihood of GE corn entering the food supply of African countries for several reasons, including the fact that current bioengineered varieties of corn are not well adapted to African growing conditions. South Africa is the only African country to commercialize biotech crops.

The Food and Agriculture Organization (FAO) of the United Nations has also offered a qualified endorsement of agricultural biotechnology, stating that it "can benefit the poor when appropriate innovations are developed and when poor farmers in poor countries have access to them.... Thus far, these conditions are only being met in a handful of developing countries." Biotechnology research and development should complement other agricultural improvements that give priority to the problems of the poor, FAO said, adding: "Regulatory procedures should be strengthened and rationalized to ensure that the environment and public health are protected and that the process is transparent, predictable and science-based."[55] Other groups have been more pointed in criticizing GE crops, arguing that they can have hidden costs that are inadequately examined by biotechnology advocates.[56]

The Global Seed Industry

Investments in crop variety research, including plant breeding, since the 1930s have led to tremendous growth in agricultural productivity. In the United States, corn productivity increased from 20 bushels per acre in 1930 to 140 bushels by the mid-1990s, a 600% increase. Much of this growth arose from innovations in major crop seeds, for example, hybridization. Such innovations have also led to further specialization in agriculture as commercial firms marketed elite crop lines developed in public land grant universities and agriculture experiment stations. This early specialization between the public and private research and development sectors has continued and accelerated, especially with innovations in plant genetic engineering and subsequent changes in intellectual property rights protection. Over the past 25-30 years, crop variety research and development has moved from a predominantly public activity to a largely private sector. Between 1960 and 1996, private investment in crop varieties increased 14-fold in inflation-adjusted dollars, while public expenditure remained largely unchanged.[57]

Seeds have distinct production processes and markets, but plant breeding, including genetic engineering and other biotechnologies, is the foundation of the seed industry today. Large, vertically integrated corporations now dominate the research and development, distribution, and marketing of seed varieties, although smaller companies may still operate under licensing and marketing agreements with the large firms. A new seed variety is typically contracted out to farmers and/or private firms for production and multiplication. Breeders provide growers with registered seed for large-scale production. Registered seed is then used to produce certified seed, which is sold commercially to farmers. Breeders closely manage contract growers to ensure that the desirable plant characteristics are retained in future seed generations.

Certified seed is dried, cleaned, sorted, and packaged ("conditioned") for sale. Under various state programs, the seed may be subjected to inspection to ensure quality. Large seed firms then market the seed to national and international markets. These firms may also license marketing to private firms to increase access to local markets. Distribution channels can differ depending on the region where the seed is marketed. For example, in the Midwest, most corn seed is sold through farmer-dealers trained by the seed company. In the South, corn seed is largely sold through agricultural supply firms. In addition, a large seed firm may sell directly to large farming operations.

The Structure of the Seed Industry

The global agribusiness sector has been undergoing consolidation and concentration for some time now. Through divestitures, mergers, and acquisitions, a few major integrated corporations currently dominate much of the agricultural input sector, (e.g., agricultural chemicals, seeds, and biotechnology traits). With the emergence of innovations in plant genetic engineering in the early 1980s, an upsurge of takeovers and mergers began within the seed industry. Chemical and pharmaceutical industries were the major purchasers of independent seed companies By 2005, according to Phillipps McDougall, a UK agribusiness consulting firm, the top 10 companies were estimated to comprise 51% of the world's commercial seed sales.[58] This figure is based on a 2005 global seed market of $19.0 billion. A smaller group of transnational firms—Monsanto, DuPont/Pioneer, Syngenta—are the industry leaders today. Between 2004 and 2005, there was an increase in seed industry acquisitions. Monsanto, through its acquisition of Seminis in 2005, displaced Dupont/Pioneer as the world's largest seed corporation (**Table 2**).

Table 2. World's Largest Seed Companies

Company	Seed Sales (Millions)
Monsanto (United States)	$6,700[a]
Dupont/Pioneer (United States)	$6,400
Syngenta (Switzerland)	$3,850
Group Limagrain (France)	$3,400
Land O'Lakes (United States)	$756
KWS AG (Germany)	$615
Bayer Crop Science (Germany)	$430

Source: ETC Group, Communique #99: Patenting the "Climate Genes" ... and Capturing the Climate Agenda." May/June 2008.

a. Includes sales from Seminis, acquired by Monsanto in 2005, and Delta and Pine Land, acquired in 2007.

Determining whether concentration and consolidation in the seed industry have reached a point where anti-competitive behavior becomes a concern requires accurate data on market share of individual firms and the total market value of the industry. Estimates for the size of the global seed market are not precise. According to one estimate, the 2006 global value of the commercial seed market was $22.9 billion. The International Service for the Acquisition of Agri-Biotech Applications (ISAAA) estimated the 2005 global market at $30 billion. In 2005, the International Seed Federation estimated the size of the market of seed and "other planting material" in 56 counties at $25.2 billion. The ETC Group estimated the total to be $21 billion. Assuming a global seed market value of $21 billion, the top 10 firms (**Table 2**) dominated approximately 49% of the market in 2004-2005.

The Herfindahl-Hirschman Index (HHI) is a measure of the size of firms relative to the overall industry and is an indicator of the degree of competition among them. As such, it is a simple measure of the degree to which a given market can be said to be competitive. The HHI is defined as the sum of the squares of the market shares of each individual firm. As the HHI decreases, it indicates a reduction in firm pricing power and an increase in competition. As the HHI increases, a few firms' pricing power increases with a corresponding decline in market competition for the particular industrial sector.

The Antitrust Division of the U.S. Department of Justice uses the HMI to determine whether a proposed merger would be anticompetitive (i.e., increase the pricing power of a few firms within a sector). The HHI for the top 4 global seed firms is 393. The Antitrust Division considers an HHI of 1000-1800 to be moderately concentrated. An HHI of 1800 or more is a concentrated industry.

Based on of the HHI, the global seed industry appears highly competitive. The simple HHI, however, can understate within-group concentration. While Monsanto, the largest seed company, has approximately 13%-14% of the global seed market, it has far greater dominance in particular seed categories. Monsanto's Roundup Ready cotton, soybeans, and canola, for example, dominate the world's genetically engineered (GE) crops, which have become an increasing share of global crop production.[59] In 2004, Monsanto's GE seed and/or its patented trait technology accounted for 175.7 million acres, approximately 88% of the total global GE crop area.[60] Monsanto has 41% of the global GE corn seed and 25% of global GE soybean seed sales.[61] **Table 3** shows Monsanto's dominance of GE commodity crops.

In addition to these bulk commodities, Monsanto has, with its 2005 acquisition of Seminis, become a dominant force in the vegetable seed market. Seminis supplies over 3,500 seed varieties to fruit and vegetable growers in 150 countries (**Table 4**).

Table 3. Monsanto's Global Share of Genetically Engineered Crops

Crop	Total Global GE Acreage	Monsanto Share of Global GE Acreage
GE Soybeans	119.5 million acres	91%
GE Corn	47.7 million acres	97%
GE Cotton	22.2 million acres	63.6%
GE Canola	10.6 million acres	59%

Source: ETC Group, Global Seed Industry Concentration -2005

Table 4. Monsanto's Share of Global Vegetable Seed Markets

Vegetable Crop	Global Share of Seed Market
Beans	31%
Cucumber	38%
Hot Pepper	34%
Sweet Pepper	29%
Tomato	23%
Onion	25%

Source: ETC Group, Global Seed Industry Concentration -2005

Monsanto also announced in August 2006 its intention to buy Delta and Pine Land (D&PL), the world's largest seed cotton company. Together, DP&L and Monsanto account for 57% of the United States cotton seed industry.[62] DP&L has subsidiaries in 13 countries, including such major cotton producers as China, India, and Brazil. This proposed merger is under scrutiny by the Antitrust Division of the U.S. Department of Justice.

ISAAA estimated the 2006 global market value of GE seeds was approximately $6.3 billion, 21% of the estimated global seed market of approximately $30 billion.[63] Estimates of market value were based on the sale price of GE seed plus any applicable technology fees. ISAAA's estimate of the distribution of GE crop market value was:

- $2.68 billion for GE soybeans;
- $2.39 billion for GE corn;
- $0.87 billion for GE cotton;
- $0.21 billion for GE canola.

Anticompetitive Behavior

Anticompetitive practices are business or government actions that prevent or reduce competition in particular markets. Such practices may include the creation of barriers to entry for firms, dumping of products on markets below their cost of production, price fixing, linking products together to limit consumer choice, government-granted monopolies, and other business actions. Anticompetitive practices are argued to have negative effects on markets and, by extension, whole economies, through the creation of monopoly profits. The assumption is that a free and efficiently functioning market economy arises when many enterprises, each with limited market power, are permitted to buy and sell. Such markets are then assumed to produce lower prices to consumers as well as a wider range of products.

Some licensing practices and conditions pertaining to intellectual property rights may also restrain competition and have adverse effects on trade and impede the transfer and dissemination of technology. Licensing practices or conditions that in particular cases constitute an abuse of intellectual property rights can have an adverse effect on competition in the relevant market.

Governments enact competition laws to prevent these and other anticompetitive practices. The realities of modern competitive markets, however, are arguably sometimes more complex than simple theories of open market competition would suggest. Oligopolistic or quasi-monopolistic firms, for example, can achieve scale economies in production or marketing that would be difficult or impossible for smaller firms to accomplish. In these production sectors (e.g., airlines), the levels of capital investment are very high, and the firms' evolution into quasi-monopolies can be an effective strategy from the standpoint of a competitive economy.

This market dominance of GE seed noted above is reinforced by Monsanto's U.S. dominance of glyphosate herbicide. In September 2006, a class-action suit involving 100,000 farmers was filed against Monsanto, the world's largest seed company, in the U.S. District Court in Wilmington, DE.[64] Plaintiffs alleged that Monsanto, through its control of 80% of the U.S. market for glyphosate, had an effective monopoly. Monsanto has patents on seed lines of cotton, canola, and soybeans that are genetically engineered to be glyphosate tolerant. In their suit, the plaintiffs alleged that Monsanto retained product exclusivity "by acquiring seed companies that were developing modified seed technology, eliminating those products that could have led to the development of genetically modified seeds that could be used with non-glyphosate herbicide." These efforts to block the development of competing genetically modified seeds had a direct effect on Monsanto's glyphosate herbicide monopoly because had competing seeds been developed, farmers would have had a choice not only to buy competing seeds, but also to use different types of herbicides instead of glyphosate. Monsanto defeated the plaintiffs' motion for class certification in July 2007; the case was dismissed without prejudice, and was refiled in the Missouri courts.

The U.S. Department of Justice and USDA have announced dates and locations of joint public workshops that will explore competition and regulatory issues in the agricultural industry. The workshops, announced in August 2009 by Attorney General Eric Holder and Secretary of Agriculture Tom Vilsack, are the first joint USDA/Department of Justice workshops ever held to explore competition and regulatory issues in agriculture. Among the topics to be discussed is that of seed concentration.

Contracts between Seed Companies and Farmers

Some criticism has arisen regarding the licensing arrangements between the corporate owners of seeds with novel traits and the farmers who plant the seeds. Because these licensing practices stem from the firms' proprietary rights on the seed germplasm, those firms with large numbers of patents are often singled out for particular attention. While all firms have technology agreements with farmers who plant their seed, Monsanto has come under particular scrutiny for its technology agreement. Not only is Monsanto the world's largest seed company, it also has the largest number of plant biotechnology patents (647 in 2003). Monsanto's Technology Agreement gives the company significant control of the patented seed after the farmer has purchased, planted, and harvested the crop. If farmers fail to follow the terms of the agreement, they can be, and often are, the subject of enforcement actions brought by Monsanto. Even farmers who claim not to have knowingly planted genetically engineered seed can suffer financial penalties when their seed is found to contain Monsanto's patented genetic material. Although seed source contamination is common, farmers who are discovered with Monsanto's patented genetic material may find themselves in legal jeopardy.[65] While these actions on the part of Monsanto have been the subject of much discussion and criticism, to date, they have not been challenged in the courts.

OTHER SELECTED ISSUES

Food Safety and Labeling

In the United States, many consumers may be wary of GE foods out of fear that introduced genes could prove allergenic, introduce increased toxicity, or otherwise be harmful to human health. Some critics express concern that FDA is placing all the responsibility on manufacturers to generate safety data, as it does normally under its pre-market approval system, and is reviewing only the conclusions of industry-sponsored studies, rather than conducting its own tests. They also believe that the process lacks transparency and adequate public scrutiny of data. Others defend the current system. They counter that additional testing and oversight are unnecessary because all foods must meet the same rigorous federal safety standards regardless of whether or not they are genetically engineered.

In July 2004, the Institute of Medicine and the National Research Council (IOM/NRC) of the National Academies of Science released a report generally supporting the proponents' view. The IOM/NRC found that food safety should be assessed based on the composition of the altered food (e.g., whether it contains new compounds, unusually high levels of nutrients, or other significant traits) rather than how the food was produced (by genetic engineering or conventional methods). However, the IOM/NRC determined that the safety of modified foods should be assessed on a case-by-case basis and cautioned that scientists' current ability to predict adverse consequences of genetic changes is limited.[66]

U.S. policy also does not require GE-derived foods to be so labeled as long as they are substantially the same as their more conventional counterparts. Nonetheless, some consumer groups continue to seek mandatory labeling of all GE foods. These groups argue that U.S. consumers, like their EU counterparts, should have an opportunity to see all relevant information on a label so that they can make food choices based on their own views about its perceived quality or safety. The food industry generally opposes compulsory labeling. It contends that consumers might interpret GE labels as "warning labels" implying that the foods are less safe or nutritious than conventional foods, when the industry believes the preponderance of science indicates otherwise. The industry also has asserted that mandatory labeling would require development of a costly and possibly unattainable system to ensure that GE and non-GE foods remain segregated from the farm to the store, with no added benefit to the consumer. The industry has asserted that if consumers want to purchase GE-free products, the market will support a voluntary system, as exists for organic foods (where rules already prohibit GE foods from being called "organic").[67]

At the international level, the Codex Committee on Food Labeling in May 2006 agreed to continue work on draft guidelines for biotech labeling, which has been under discussion for approximately 10 years. Committee members asked a work group co-chaired by Norway, Argentina, and Ghana to examine member countries' biotech labeling policies, their rationale, and experiences, among other questions.[68] Over the objections of the United States, Canada, and several Latin American countries, the Codex Committee will continue work on biotechnology labeling. At an April 2008 meeting of Codex in Ottawa, the EU, most Asian nations, and African countries defended their mandatory labeling regimes as necessary to protect consumers.[69]

Adventitious Presence

A related question is the definition of "mixing" and whether there should be a threshold *de minimis* amount of GE material permissible in non-GE material. "Adventitious presence" (AP) refers to any incidental appearance of very small amounts of foreign material in a commodity, food, or feedstuff. This can occur at any time during production, harvesting, storage, or marketing. Another related question is how to assess liability if such mixing does occur, or if GE plants prove harmful to the environment. For example, to what extent if any should biotechnology companies share liability with producers and others who use their products?

Presently in the grain business, even shipments of the highest grades are permitted to contain some specified low levels of unwanted material, such as weeds, damaged kernels, and/or stems and leaves. Corn graded No. 1, for example, may contain up to 2% foreign material. As more crops and acreage are devoted to GE varieties, it becomes increasingly difficult, if not impossible, to avoid their trace presence in non-GE varieties.

No internationally recognized standards have existed for what amounts, if any, of GE material should be permitted in a non-GE crop, especially if that crop or a food derived from it will be labeled as non-GE. In the absence of such standards, individual countries are establishing their own, often varying, AP thresholds. The lack of consistent, scientifically sound standards is confusing consumers and disrupting trade, the biotech industry has asserted. For example, the new EU regulation sets a tolerance level for non-GM foods, feeds, and processed products at 0.9%. All products with more than 0.9% must be labeled as GM. U.S. agricultural interests consider the EU regulation in particular to be unworkable and discriminatory. EU officials counter that their standards not only are reasonable but also are being demanded by consumers. (See also "U.S.-EU Dispute," above.)

In its January 23, 2004, notice, APHIS asked for comments on if, and how, its regulations should address the LLP question for GE plant material. Questions include whether such presence should be exempt from regulation, what thresholds (levels) of low-level presence (LLP) might be acceptable, and under what conditions. Major grain and biotechnology industry organizations responded by urging the FDA, EPA and APHIS to establish a policy governing LLP. In March 2007, APHIS published a *Federal Register* notice describing how the agency responds when LLP of regulated GE materials occur in commercial seed or grain that may be used for food or feed.[70] In the proposed APHIS regulation revisions discussed above, APHIS has proposed establishing criteria under which the occurrence of LLP may not be cause for agency remedial action. The new provision would permit APHIS to determine that a LLP event is non- actionable when the criteria support the conclusion that the LLP is unlikely to result in the introduction or dissemination of a plant pest or noxious weed.

Environmental Concerns

Two main issues continue to drive the science and public debate on the environmental impacts of GMOs. One issue is the transfer of the introduced genes to wild plants and non-GM crops (i.e., gene flow from GE plants). The second issue concerns the indirect effects of the GE crops themselves on the local environment. Aside from the contamination issue to producers who do not want to plant GE crops, there is the concern that introduced genes can lead to herbicide and pesticide resistance in non-target species.[71] There is mounting evidence that some weeds have begun to show resistance to glyphosate/Roundup. Several varieties of rye grass, Palmer amaranth (pigweed), common waterhemp, and giant ragweed are showing signs of resistance to glyphosate, in addition to the common ragweed.[72]

Biotechnology advocates claim that GE crops offer environmental advantages over conventionally produced organisms. They note that the technology is more precise than traditional methods like crossbreeding. The latter methods transfer unwanted and unanticipated characteristics along with the desired new traits from one organism to another. Biotechnology also has made it possible to apply fewer and less toxic chemical herbicides and insecticides and to reduce soil tillage (thereby decreasing erosion and improving soil fertility), supporters of the technology assert.

Critics counter that genetic engineering is not like traditional breeding. It creates crop and animal varieties that would not otherwise occur in nature, posing unpredictable risks to the environment (and to human health), they point out. Because they are living organisms, GE crops are difficult to control, greatly increasing the potential for escaping into the environment, crossbreeding with and overtaking wild species, and generally disrupting the natural ecosystem, critics believe. For example, GE, herbicide-tolerant seeds or pollen could inadvertently create "superweeds" that out- compete cultivated or wild plants, critics argue.

A 2002 NAS/NRC report stated that it could find no new distinctions between the types of environmental risks posed by GE plants and those posed by more conventionally bred crops (and that, in fact, there is a need to re-evaluate the potential environmental effects of the latter). The study concluded that the current APHIS regulatory system for biotechnology had improved substantially since it was first initiated and is more rigorous than the environmental oversight for other agricultural products and practices. The study did fmd areas of concern, including the need for greater transparency and public input into the regulatory process, and for more ecological monitoring after GE plants are approved and enter the marketplace.

A 2004 NAS/NRC report cited studies to conclude that some GE organisms are viable in natural ecosystems and can breed with wild relatives. The report urged developers of GE organisms to consider biological techniques such as induced sterility in order to prevent transgenic plants and animals from escaping into the environment. "Because no single bioconfinement method is likely to be 100% effective," and because few are well-developed, such developers should create a redundant system by using more than one method of containment. The report called for more research to improve both containment methods and public confidence in regulation.[73] In May 2004, a separate report by University of Arizona and Texas A&M University researchers confirmed the spread of GE corn into a nearby field of non-GE corn.[74] In September 2004, a team of researchers from the Environmental Protection Agency confirmed the spread of GE grass pollen to non-GE grass up to 13 miles away, much further than previous studies would have indicated.[75]

Plant-Based Pharmaceuticals from Biotechnology

Worldwide, hundreds of GE plants are under development for use as "factories" for pharmaceuticals (and other industrial compounds). Between 2004-2007 approximately 485 acres in the United States were planted to regulated GE plants for filed testing of plants producing pharmaceuticals, industrial compounds, and value-added chemicals for human consumption or phytoremediation.[76] None of these compounds has been commercialized to date. Pharmaceuticals might include, for example, vaccines or medicines for forms of cancer, infectious diseases, cardiovascular and nervous system diseases, metabolic disorders, and agents of biowarfare.

A National Research Council Report in 2004 recognized that "biopharm crops pose a wholly different order" of environmental and human health risks.[77] APHIS announced in 2007 that an environmental impact statement was being prepared for field trials of a transgenic sunflower that in engineered to produce human proinsulin, which tests have shown to be structurally, chemically, and functionally the same as pharmaceutical grade human insulin

Proponents believe plant-based pharmaceuticals will provide a far more cost-effective alternative to conventional pharmaceutical production, which now requires major investments both in large volumes of purified culture mediums and in manufacturing plants. Plant-based pharmaceuticals, on the other hand, may be more easily incorporated into the existing agricultural infrastructure, providing a significant new source of farm income, they believe.[78]

Critics are concerned about impacts on the food supply if crops like corn (the most widely planted U.S. crop, an intensively researched plant for biotechnology, and also an airborne pollinator) are "pharmed." In 2002, for example, material from GE-altered corn plants that had been test-planted in a prior growing season in Nebraska for pharmaceutical use (for ProdiGene, Inc.) was inadvertently mixed with some 500,000 bushels of soybeans, which had to be quarantined by USDA to keep them out of the food supply. USDA officials observed that the soybeans never reached the food or feed supply, evidence that current regulatory oversight is effective. Some critics argue that GE plants producing pharmaceuticals and industrial compounds should be evaluated by criteria different from those used to evaluate crops intended for food. Others have argued that biopharm plants should not be food crops.

Concerns persist among both consumer groups and the food manufacturing industry about producing GE plant-made pharmaceuticals in food crops. Some want 100% prevention systems in place before the first product is commercialized. Some of these groups suggest that only non-food crops should be used for GE plant-made pharmaceuticals, or that, at a minimum, pharmaceutical crops should be banned from agricultural areas where food and feed crops are produced. Other potential issues include whether manufacturers of plant-based pharmaceuticals will be able to maintain consistency in dosages and overall quality, and unanticipated environmental problems (e.g., threatening endangered species).[79]

Responding to such concerns, APHIS published in the March 10, 2003, *Federal Register* a notice tightening permit conditions for its 2003 field tests of GE plants with pharmaceutical and industrial traits. The changes included (1) doubling the minimum distance allowed between traditional corn fields and test sites of pharmaceutical or industrial corn; (2) for all pharmaceutical crops (corn and other), doubling fallow zones around test sites; (3) restricting what can be grown on a test site and fallow zone in the next growing season; (4) using dedicated machinery (e.g., harvesters, planters) and storage facilities only for pharmaceutical production—adequate cleaning for other uses is no longer acceptable; (5) submitting for APHIS approval equipment cleaning and seed cleaning and drying procedures; (6) increasing APHIS field site inspections from one per season to five per season plus two visits the following year to look for any volunteer plants; (7) more record-keeping and training requirements. APHIS issued a letter on January 14, 2004, aimed at clarifying and updating its previous guidance on permits.[80] The proposed APHIS revisions for regulating GE plants discussed above would put GE plants expressing pharmaceuticals or industrial compounds in a risk category where the engineered trait had a high potential for harm. It is not the highest risk category. APHIS equated the biopharmed plants with a poplar engineered to produce enzymes for heavy metal remediation.[81]

In early August 2006, a U.S. district court judge in Hawaii ruled that APHIS had violated the federal Endangered Species Act (P.L. 93-205) and the National Environmental Policy Act (P.L. 91-190) because it had failed to consider potential impacts on endangered species and critical habitats prior to approving field trials for pharmaceutical corn on more than 800 acres throughout the Hawaiian Islands. The four companies issued the permits by APHIS were ProdiGene, Monsanto, Hawaii Agriculture Research Center and Garst Seed. All of the companies' plants used to make pharmaceutical crops had been harvested before the suit was filed and the companies stopped planting the crops under the permits. Spokesmen for both Syngenta, which subsequently bought Garst, and Monsanto, said at the time they no longer intend to pursue research into making drugs from plant crops.

EMERGING POLICY ISSUES

In the coming decade, several policy issues are likely to be at the center of attention by industry, consumer groups, and policy makers. From a general perspective, some of the issues revolve around managing the coexistence of traditional agricultural production with the increased presence of GE-based agricultural production. In some respects, the policy and regulatory issues may not be fundamentally new or different from the biotechnology issues of the past twenty years. Rather, certain issues may increase in importance as the industry matures and these longer- standing regulatory issues take new forms. While not exhaustive, some of these issues might include:

- evolving technologies, including the introduction of new "stacked trait" varieties—plant varieties with multiple genetically engineered traits—which is likely to increase; continuing development and the eventual commercialization of GE plant-based industrial and pharmaceutical output traits; oversight of second- generation biotechnology traits such as improved nutritional qualities and resistance to environmental stress (e.g., drought);
- transgenic animals and the food and industrial/pharmaceutical products derived from them; animal welfare concerns;
- importation of GE products;
- low-level presence of unapproved GE materials in food and feed products;
- the adequacy of environmental impact assessments;

- issues related to transparency and the participation in policy and regulatory issues by various stakeholders (e.g., consumers, religious groups, animal welfare activists);
- compliance with existing and emerging regulatory structures in the United States and our trading partners; testing and measurement issues; traceability and labeling of GE products.

As noted above, the evolution of herbicide-resistant weeds, especially those resistant to glyphosate, is a growing concern. As herbicide resistance increases among weed varieties, there could be increased reliance on herbicides that are less benign than glyphosate (e.g., dicamba, 2,4 D). Biotechnology companies have begun engineering new plant varieties that are tolerant to these herbicides, or varieties where several herbicide-tolerant traits are "stacked" into a single variety.[82] The environmental effects of the increasing herbicide resistance and the resort to other herbicides may become policy issues as companies seek to commercialize these new varieties.

In Congress

Congress generally has been supportive of GE products, although some Members have expressed wariness about their adoption and concerns about how they are regulated. Over the past several years, legislative activity has been relatively subdued. Congress continues to fund a variety of biotechnology-related activities at USDA, primarily through regular annual appropriations. Most of the USDA spending for biotechnology related programs is for various types of research (mainly through the Department's Agricultural Research Service and the Cooperative State Research, Education, and Extension Service). APHIS's BRS budget in FY2009 was $16.4 million This was approximately $5 million more than in FY2008. Most of the increase supported APHIS's Biotechnology Quality Management System, a new voluntary, audit-based compliance assistance program. For FY2010, the BRS budget was $13 0 million, and for FY2011 the estimated budget is $18.7 million Other USDA agencies also have received lesser amounts for various biotechnology activities.

Animal cloning, although not involving transgenic animals to date, has been an area of some interest. **In the 111th Congress**, Representative Fortenberry introduced H.R. 110, the Human Cloning Prohibition Act of 2009. Representative Kucinich introduced H.R. 5578, the Genetically Engineered Safety Act, which would prohibit open-air cultivation of GE pharmaceutical and industrial crops and the use of common human food or animal feed as the host plant for a genetically engineered pharmaceutical or industrial chemical. Representative Kucinich also introduced H.R. 5579, the Genetically Engineered Technology Farmer Protection Act. The bill would provide various legal protections for farmers and ranchers that may be harmed economically by GE seeds. The bill also would assign liability for injury caused by GE organisms.

In the 110th Congress, Representative Kucinich also introduced the same bills he introduced in the 111th Congress (1) H.R. 6635, Genetically Engineered Safety Act; and (3) H.R. 6637, Genetically Engineered Technology Farmer Protection Act. He also introduced H.R. 6636, Genetically Engineered Food Right to Know Act, would require foods that contained GE material to be so labeled.

In the 109th Congress, members continued to follow trade developments, particularly the U.S.EU dispute, and GE rice case, as well as U.S. regulatory mechanisms for approving biotech foods. However, there were few proposed bills. In May 2006, Representative Kucinich again introduced a series of bills, like those he offered in the 108th Congress, to provide what he called "a comprehensive regulatory framework" for GE plants, animals, and other organisms. The bills were H.R. 5266, H.R. 5267, H.R. 5268, H.R. 5269, H.R. 5270, and H.R. 5271.

In the 108th Congress, after the Administration launched a formal challenge of the EU GMO moratorium, the Senate on May 23, 2003, passed by unanimous consent a resolution (S.Res. 154) in support of the action. A similar House measure (H.Res. 252) was passed on June 10, 2003, by a suspension vote of 339-80. Also in the 108th Congress, Representative Nick Smith introduced bills (H.R. 2447, H.R. 3472, H.R. 4651) to create an interagency task force to promote the benefits of agricultural biotechnology. Both bills were referred to the House Agriculture Committee, but no subsequent action was taken on them. Other members took a different approach in proposing bills related to food and agricultural biotechnology.

Representative Kucinich introduced a series of bills during the 108th Congress (H.R. 2916, H.R. 2917, H.R. 2918, H.R. 2919, H.R. 2920, H.R. 2921) that would have prescribed a variety of legislative changes intended to mandate labeling of GE-based foods, broaden FDA oversight, protect producers from any potential legal and environmental risks from agricultural biotechnology, prohibit unapproved U.S. exports of GE plants and animals, and tighten rules for producing and handling GE pharmaceutical and industrial crops, among other things. Senator Durbin introduced a bill, S. 2546, to require premarket consultation and approval for GE foods at the FDA. These bills were referred to various committees, but no further action was taken on them by the 108th Congress.

ACKNOWLEDGMENTS

This report was originally written in collaboration with Geoffrey S. Becker, Specialist in Agricultural Policy.

End Notes

[1] Among the sources for this report are various materials by USDA's Economic Research Service (ERS) and Animal and Plant Health Inspection Service (APHIS), the Pew Initiative on Food and Biotechnology, various issues of *Food Chemical News,* a weekly trade publication, and the Biotechnology Industry Organization (BIO).

[2] OECD. *Policy Brief, Modern Biotechnology and the OECD,* June 1999. http://www.oecd.org/dataoecd/29/40/1890904.pdf.

[3] FAO. *FAO Biotechnology Glossary,* http://www.fao.org/biotech/indexglossary.asp.

[4] International Service for the Acquisition of Agri-biotech Applications (ISAAA), ISAAA Brief 39-2008, *Executive Summary: The First Thirteen Years, 1996 to 2008 Report on the Global Status of Biotech/GM Crops.* Accessed January 2008 at http://www.isaaa.org/resources

[5] Sources: Information Systems for Biotechnology at Virginia Tech; also, USDA, ERS, *The First Decade of Genetically Engineered Crops in the United States,* April 2006, which can be accessed at http://www.ers.usda.gov/ Publications/eib11/.

[6] Because Bt is a natural occurring pesticide, it can be used under certain conditions on organically produced plants. Its incorporation into GE commodities concerns some organic producers because of the risk of creating Bt tolerant pests thereby decreasing the utility of Bt to organic farming operations.

[7] Some of the reduced public opposition to the GE beets may be based on the fact that sugar crystals do not contain any remnants of the GE modified protein and, thus, could pose no dietary risk.

[8] The legal challenge seeking an injunction against was filed in the U.S. District Court for the Northern District of California and also includes the Sierra Club and the Organic Seed Alliance as plaintiffs. The court filing may be accessed at http://www.centerforfoodsafety.org/pubs/Final%20Complaint.pdf.

[9] ERS, *The First Decade of Genetically Engineered Crops in the United States.* April, 2006. http://www.ers.usda.gov/ publications/eibll/eib 1 1 .pdf

[10] Milk containing rBST may be falling out of favor in some places. Wal-Mart, the largest grocery retail outlet in the United States, has a private label milk (Great Value Milk) that will be rBST free. Kroger completed a phase-out of rBST milk in February 2008. Safeway switched to rBST free milk in its private line, although it continues to sell other rBST milk. Starbucks began using only rBST free milk in January 2008.

[11] Also see CRS Report RL33334, *Biotechnology in Animal Agriculture: Status and Current Issues,* by Tadlock Cowan.

[12] "FDA approves drug product from transgenic goats." *Food Chemical News,* February 9, 2009.

[13] In September 2008, APHIS announced a request for public comment and technical empirical data concerning ongoing and future research on genetically engineered animals.

[14] Sources include Cornell University, Genetically Engineered Organisms Public Issues Education Project (GEO-PIE), at http://www.geo-pie.cornell.edu/crops accessed on February 3, 2009; USDA, APHIS, Petitions of Nonregulated Status Granted or Pending by APHIS, at http://www.aphis.usda.gov/brs/not_reg.html and Colorado State University, *Transgenic Crops: An Introduction and Resource Guide.* The latter report is available at http://www.colostate.edu/programs/lifesciences/TransgenicCrops/index.html. The site is not regularly maintained but archived materials are available through 2004.

[15] Sources include "Review of Agricultural Biotechnology," hearing before the Subcommittee on Conservation, Credit, Rural Development, and Research of the U.S. House Committee on Agriculture, June 23, 2004 (Serial No. 108-34); BIO; Colorado State University; and ERS, *Economic Issues in Agricultural Biotechnology* (AIB-762), February 2001 (table, p. 19), at http://www.ers.usda.gov/publications/aib762/; and *The First Decade of Genetically Engineered Crops in the United States.*

[16] So *far* one GE fish, the "Glofish," has been marketed in the United States. It is an aquarium fish that is not approved for consumption. For more on genetically engineered fish, see CRS Report RL32974, *Genetically Engineered Fish and Seafood,* by Rachel Borgatti and Eugene H. Buck.

[17] See the FDA biotechnology website at http://www.cfsan.fda.govt--Ird/biocon.html#policy.

[18] FDA's Recommendations for the Early Food Safety Evaluation of New Non-Pesticidal Proteins Produced by New Plant Varieties Intended for Food Use can be accessed at http://www.cfsan.fda.gov/~dms/bioprgu2.html. The guidance was issued in draft form in November 2004 and had earlier been proposed by OSTP in 2002.

[19] See CRS Report RL33334, *Biotechnology in Animal Agriculture: Status and Current Issues,* by Tadlock Cowan.

[20] These studies include the Institute of Medicine/National Research Council 2004 report *Safety of Genetically Engineered Foods: Approaches to Assessing Unintended Health Effects;* the National Academy of Sciences/National Research Council (NAS/NRC) 2002 report *Environmental Effects of Transgenic Plants: The Scope and Adequacy of Regulation;* the NAS/NRC 2000 report *Genetically Modified Pest-Protected Plants: Science and Regulation;* the Council for Agricultural Science and Technology (CAST) 2001 report *Evaluation of the U.S. Regulatory Process for Crops Developed Through Biotechnology;* and the CAST 2002 report *Comparative Environmental Impacts of Biotechnology-derived and Traditional Soybean, Corn, and Cotton Crops.*

[21] In January 2008, APHIS announced the preparation of an environmental impact statement on GE alfalfa. See *Federal Register,* Vol. 73, No. 4, January 7, 2008: 1198-1200

[22] For a discussion of the technical issues in growing GE alfalfa in proximity to non-GE alfalfa, see the following University of California-Davis study: http://alfalfa.ucdavis.edu/2007AlfalfaConference/2007/07-96.pdf.

[23] USDA Advisory Committee on Biotechnology and 21st Century Agriculture. *Opportunities and Challenges in Agricultural Biotechnology: The Decade Ahead.* July 13, 2006. The committee consists of biotech industry, agricultural, consumer and scientific representatives. Accessed January 2009 at http://www.usda.gov/wps/ portal/!ut/p/ J.7_0_A/7_0_10B?contentidonly=true&contentid=AC21Reports.xml.

[24] The various arguments are explored in more depth in an April 2004 Pew Initiative report, *Issues in the Regulation of Genetically Engineered Plants and Animals.* See http://pewagbiotech.org/.

[25] As reported in "FDA issues 'adventitious presence guidance for biotech plants," in *Food Chemical News,* June 26, 2006. See page 17 for additional discussion of the AP issue.

[26] A copy of the lawsuit and an accompanying press release can be viewed at the CFS website at http ://www. centerforfoodsafety. org/Ge_Foods_FDA. cfm.

[27] USDA, OIG. *Animal and Plant Health Inspection Service Controls Over Issuance of Genetically Engineered Organism Release Permits,* accessed January 2009 at http://www.usda.gov/oig/webdocs/50601-08-TE.pdf.

[28] "BRS OIG Report Frequently Ask Questions," which responds in more detail to the OIG criticisms. It can be viewed at http://www.aphis.usda.gov/brs/brs_oig.html.

[29] USDA, Office of the Inspector General, Southwest Region, *USDA Controls Over Importation of Transgenic Plants and Animals,* Report No. 50601-17-Te, December 2008.

[30] *Federal Register.* Vol. 73, No. 197, 60009-60048, October 9, 2008.

[31] Only a small fraction of weeds are considered to be noxious weeds. APHIS currently lists 98 aquatic, terrestrial, or parasitic plant taxa as noxious weeds.

[32] "Industry, NGOs, strongly oppose proposed USDA biotech regulation." *Inside U.S. Trade,* October 17, 2008. .

[33] Food Chemical News. "USDA Proposed Biotech Regulatory Overhaul to Mixed Reviews," vol. 50, no. 34, October 13, 2008.

[34] Alfalfa is an open-pollinated plant. Non-GE alfalfa would be at risk of being pollinated through windborne pollen from GE alfalfa. The impact on non-GE alfalfa growers was deemed a significant environmental impact.

[35] The draft EIS may be downloaded at APHIS's website: http://www.aphis.usda.gov/biotechnology gealfalfa_deis.pdf.

[36] *Monsanto Co. v. Geerston Seed Farms,* 130 S. Ct. 2743 (2010)

[37] Monsanto Company and KWS SAAT AG Petition 03-323-01p for Determination of Nonregulated Status for Roundup-Ready Sugar Beet Event H7-1. USDA/APHIS Environmental Assessment and Finding of No Significant Impact, February 2005. Document available at http://www.aphis.usda.gov/brs/aphisdocs2/04_11001p_com.pdf.

[38] *Center for Food Safety v. Vilsack,* No. C08-00484 JSW (N.D. Cal. 2009).

[39] *Federal Register,* vol. 74, 106, June 4, 2009, http://www.aphis.usda.gov/brs/fedregister/BRS_20090604b.pdf.

[40] In February 2008, APHIS, EPA, and FDA announced a coordinated response to a notification by Dow AgroSciences that the company had detected an unregistered GE pesticide product know as a plant-incorporated protectant (PIP) in three lines of its commercial hybrid seed lines. Concerns that plant-derived pharmaceutical and industrial products could enter the food and feed system are based on difficulties in ensuring that unapproved plants are not commingled with approved varieties.

[41] *Center for Biological Diversity, et. al v. Animal and Plant Health Inspection Service.* Case 2:10-cv-14175-KMM. Document available at http://www.biologicaldiversity.org/programs/public_lands/forests/pdfs/Eucalyptus-Complaint-signed.pdf.

[42] See also CRS Report RL31970, *U.S. Agricultural Biotechnology in Global Markets: An Introduction,* by Geoffrey S. Becker and Charles E. Hanrahan. This report does not discuss the trade challenges encountered by the biotechnology companies themselves. Among other problems, besides foreign resistance to agricultural biotechnology in general, these companies also face often divergent laws on international property rights (IPR), where their patent or plant breeding rights in one country may be nonexistent in another. In the developing world in particular, the policy challenge is to find a balance between companies' IPR and the ability to use the new technologies. For details, see International Food Policy Research Institute, *Biotechnology and Genetic Resource Policies,* Briefs 1-6, January 2003; and CRS Report RL31568, *Plants, Patents, and Seed Innovation in the Agricultural Industry,* by John R. Thomas.

[43] "Statement by Agriculture Secretary Mike Johann Regarding Genetically Engineered Rice," August 18, 2006. LL stands for "Liberty Link," a trademark name for the herbicide glyphosinate. LL crops are engineered to tolerate the herbicide, making for more effective weed control.

[44] Statement of Bill J. Reed, Riceland Foods' vice president for public affairs, August 18, 2006, as quoted by the website AgWeb.com.

[45] Center for Food Safety, "Unapproved, Genetically Engineered Rice Found in Food Supply," August 18, 2006, press release.

[46] Sources include *Food Chemical News,* various issues; Cornell University GEO-PIE; and several news wire service reports.

[47] See http://www.gmofoodforthought. com/2007/10/time_is_right_for_biotech_whea.html.

[48] The joint statement can be accessed at http://www.wheatworld.org/userfiles/file/ FINAL%20Trilateral% 20 Biotech% 20Statementpdf.

[49] *Definitive Global Rejection of Genetically Engineered Wheat.* The statement can be accessed at http://cban.ca/Resources/Topics/GE-Crops-and-Foods-Not-on-the-Market/Wheat/Definitive-Global-Rejection-of-GeneticallyEngineered-Wheat

[50] "Second Round of Fighting Begins Over Genetically Engineered Wheat," *Digital Journal,* July 17, 2009, at http://www.digitaljournal.com/article/276000.

[51] See CRS Report RS21556, *Agricultural Biotechnology: The U.S.-EU Dispute,* by Charles E. Hanrahan.

[52] Sources include CRS Report RL30594, *Biosafety Protocol for Genetically Modified Organisms: Overview,* by Alejandro E. Segarra and Susan R. Fletcher; and various USDA and U.S. State Department background materials.

[53] "China Could Be First Nation to Approve Sale of GM Rice," *Science,* 306:1458-1459 (November 26, 2004); plus various USDA agricultural attached reports. This point was also underlined by USDA's agricultural biotech advisory committee in its July 13, 2006, report.

[54] A United Nations Report recognized that biotechnology could play a positive role in development. See UN Human Development Report, *Making New Technologies Work for Development.* 2001. Report available at http://hdr.undp.org/ en/reports/global/hdr2001/

[55] Food and Agriculture Organization, *The State of Food and Agriculture 2003-2004,* at http://www.fao.org/ documents/ show_ cdr.asp?url_file=/docrep/006/y5160e/y5160e00.htm.

[56] See Friends of the Earth. *Who Benefits from GM Crops?: The Rise in Pesticide Use.* January, 2008. Report may be accessed at http://www.centerforfoodsafety.org/pubs/FoE%20I%20Who%20Benefits%202008%20- %20 Full% 20 Report% 20FINAL%202-6-08.pdf

[57] Ferdandez-Cornejo, Jorge and David Schimmelpfennig. "Have seed industry changes affected research effort?" *Amber Waves.* Economic Research Service, USDA, February 2004.

[58] Phillips McDougall, "Seed Industry Consolidation," July 2005. Unpublished report cited in United Nations Environmental Program, Convention on Biological Diversity, Compilation of Submissions on Potential Socio-economic Impacts of Genetic Use Restriction Technologies (GURTs) on Indigenous and Local Communities. UNEP/CBD/WG8J/4/INF/6, December 14, 2005.

[59] Roundup is the trademark name for Monsanto's glyphosate herbicide. GE varieties developed and patented by Monsanto are resistant to glyphosate

[60] ETC Group. *Global Seed Industry Concentration-2005.* ETC Group Communique, September/October 2005

[61] Ibid., page 5.

[62] Bayer Crop Science, a top-10 seed company, accounted for about 25% of the U.S. cotton seed market in 2005.

[63] International Service for the Acquisition of Agri-Biotech Application (ISAAA). *ISAAA Brief 35-2006: Global Status of Commercialized Biotech/GM Crops.* January 18, 2007. http://www.isaaa.org/resources executive summary/default.html. Site accessed January, 2009.

[64] Pullen Seeds and Soil v. Monsanto Company, No. 06-599,

[65] Center for Food Safety. *Monsanto vs. U.S. Farmers.* 2005. http://www.centerforfoodsafety. org/Monsantovsus farmers report.cfm. Accessed January 2009.

[66] Press release, "Composition of Altered Food Products, Not Method Used to Create Them, Should Be Basis for Federal Safety Assessment," The National Academies, July 27, 2004.

[67] An industry group—the Non-GMO Project—concerned that GMOs may be commingling with organic and "natural" foods has begun a campaign to test products and label them if they contain GMOs. Whole Foods Market, one of the participants, plans to place its seal on hundreds of products it markets under its "365" brand.

[68] "Report of the Thirty-fourth Session of the Codex Committee on Food Labeling," Ottawa, Canada, 1-5 May 2006, as presented to the Codex Alimentarius Commission Twenty-ninth Session, Geneva, Switzerland, 3-7 July 2006.

[69] *Food Chemical News.* "Codex panel continues work on biotech labeling." Vol. 50, No.11, May 5, 2008.

[70] Federal Register. Policy on Responding to the Low-Level Presence of Regulated Genetically Engineered Plant Material." Vol., 14649-14651, March 29, 2007.

[71] Australian farmers grow non-GE imadiazolone-tolerant canola and non-GE triazine-tolerant canola. It is possible that if GE glyphosate-tolerant canola and GE glufosinate ammonium-tolerant canola are introduced, contamination could result in canola plants being unintentionally bred to be resistant to all four chemicals, glyphosate, gluphosinate ammonium, triazine, and imadizolone. The canola plants would be considered a "super weed."

[72] "Glyphosate Resistant Weeds a Growing Concern," *Minnesota Farm Guide,* April 25, 2008, at http://www. minnesotafarmguide.com/articles/2008/04/25/ag_news/production_news/prol 1.txt. On July 28, 2010, the House Committee on Oversight and Government Reform's Domestic Policy Oversight Subcommittee held a hearing on the emergence of herbicide resistant weeds, with a particular focus on glyphosate-resistant weeds.

[73] NAS/NRC, respectively, Environmental Effects of Transgenic Plants: The Scope and Adequacy of Regulation, 2002; and Biological Confinement of Genetically Engineered Organisms, 2004. Among numerous other studies that examine environmental impacts and the adequacy of regulation are Council for Agricultural Science and Technology, Comparative Environmental Impacts of Biotechnology-derived and Traditional Soybean, Corn, and Cotton Crops, June 2002; and Pew Initiative on Food and Biotechnology, Post-Market Oversight of Biotech Foods—Is the System Prepared? (prepared for Pew by Resources for the Future), April 2003.

[74] "Contamination of refuges by Bacillus thuringiensis toxin genes from transgenic maize," Charles F. Chilcutt and Bruce E. Tabashnik, Proceedings of the National Academy of Sciences, May 18, 2004, 752-7529.

[75] Proceedings of the National Academy of Sciences, "Evidence for landscape-level, pollen-mediated gene flow from genetically modified creeping bentgrass with CP4 EPSPS as a marker," Watrud et al., at http://www.pnas.org/cgi/doi/ 10.1073/pnas.0405154101.

[76] APHIS release permits (e.g., *NEPA Documents* and *Supplement Permit Conditions")are* publicly available at http://www.aphis.usda.gov/brs/ph_permits.html.

[77] National Research Council. Biological Confinement of Genetically Engineered Organisms. Washington , D.C., 2004.

[78] Also see CRS Report RS21418, *Regulation of Plant-Based Pharmaceuticals,* by Geoffrey S. Becker and Donna U. Vogt.

[79] The 2004 NAS/NRC report observed that an organism widely used for food "probably would be a poor choice as a precursor for an industrial compound" unless it were strictly confined. Alternative nonfood host organisms should be sought, the report concluded.

[80] The latest version of this guidance *(Draft Guidance for APHIS Permits for Field Testing or Movement of Organisms with Pharmaceutical or Industrial Intent,* March 31, 2006) is available at http://www.aphis.usda.gov/brs/pdf/ Pharma_Guidance.pdf.

[81] *Federal Register.* Vol. 73, No. 197, October 9, 2008: 60008-600048.

[82] A new dicamba-resistant soybean is expected to be available as soon as 2013. See http://deltafarmpress.com/dicamba-resistant-soybeans-2013. Dow AgroScience is currently developing a 2,4 D tolerant variety of corn. See http://www.dowagro.com/newsroom/corporatenews/2010/20100303a.htm.

In: Technology Policies and Issues for the U.S. Government ISBN: 978-1-61209-501-1
Editors: Danielle A. Godden and Mason C. Andrews

Chapter 5

AGRICULTURAL BIOTECHNOLOGY: THE U.S.-EU DISPUTE *

Charles E. Hanrahan

SUMMARY

In May 2003, the United States, Canada, and Argentina initiated a dispute with the European Union concerning the EU's *de facto* moratorium on biotechnology product approvals, in place since 1998. Although the EU effectively lifted the moratorium in May 2004 by approving a genetically engineered (GE) corn variety (MON810), the three complainants pursued the case, in part because a number of EU member states continue to block already approved biotech products. Industry estimates are that the moratorium costs U.S. corn growers some $300 million in exports to the EU annually. Corn gluten exports from the United States to the EU have been blocked since 2007 because of a zero tolerance policy governing the accidental presence of non-approved U.S. GE corn in such shipments.

On November 21, 2006, the WTO's Dispute Settlement Body (DSB) adopted the dispute panel's report, which ruled that a moratorium had existed, that bans on EU-approved GE crops in six EU member countries violated WTO rules, and that the EU failed to ensure that its approval procedures were conducted without "undue delay." The EU announced it would not appeal the ruling. The United States and EU agreed on November 21, 2007 (subsequently extended to January 11, 2008), as a deadline for EU implementation of the panel report. On January 11, 2008, the U.S. Trade Representative announced that, while it was reserving its rights to retaliate, it would hold off seeking a compliance ruling while the United States sought to normalize trade in biotechnology products with the EU.

In the meantime, co-complainants Canada (July 15, 2009) and Argentina (March 18, 2010) have reached "final settlements" in the biotech dispute with the EU. Canada, Argentina, and the EU notified the DSB of their mutually agreed solution under Article 3.6 of the DSU.

---

* This is an edited, reformatted and augmented version of Congressional Research Service publication, Report RS21556, dated April 8, 2010.

The parties agreed to establish a bilateral dialogue on agricultural biotech market access issues of mutual interest.

U.S. agricultural and trade officials continue to criticize the EU for its biotech approval processes. During the second session of the 111th Congress, Members with agricultural interests may debate the issue of whether to continue a dialogue with the EU on re-establishing trade in biotechnology products or to seek retaliation for presumed lack of EU compliance with the panel decision.

BACKGROUND

In May 2003, the United States, Canada, and Argentina requested consultations—the first step in WTO dispute settlement—with the European Union (EU) concerning the latter's *de facto* moratorium since 1998 on approving new genetically engineered (GE) products. U.S. agricultural interests contended that these policies not only blocked their exports to the EU, but also fueled unwarranted concerns about the safety of agricultural biotechnology throughout the world. The United States also maintained that a number of EU member states maintained national marketing and import bans on GE products even though those products had already been approved by the EU for both marketing and import in the EU. Although the EU recommenced approvals in May 2004 with the approval of a GE corn variety for human consumption, a number of EU member states continue to block dissemination of approved biotech varieties. The EU restarted its approval process following adoption of new labeling and traceability rules for GE crops and foods in 2003.[1]

The United States accounted for 48% of the 331 million acres (134 million hectares) planted globally with GE crops in 2009.[2] In 2009, 91% of all U.S. soybean, 88% of U.S. cotton, and 85% of U.S. corn acres were planted with GE varieties, designed mainly to control pests (weeds and insects).[3] GE production is not segregated from production from conventional varieties because the U.S. regulatory system recognizes GE crops (once approved for commercialization) as substantially equivalent to non-GE varieties. Gaining market acceptance of GE crops within the United States has been easier than overseas, however, where, in markets like the EU, consumers and their governments have been more wary of biotechnology.

With minor exceptions, the EU and its member states approved no GE products between 1998 and 2004. As of January 2004, 22 GE products or crops were awaiting approval. A bloc of EU states had effectively halted the release of any new GE crops into the environment, saying that they would not implement the EU-wide legislation for approvals until new, stricter regulations for labeling and tracing GE-containing products (discussed below) took effect.[4]

In the three years before the *de facto* ban, U.S. corn exports to the EU averaged about $300 million annually, according to USDA data (Spain and Portugal were the largest EU importers). During the ban, they declined to less than one-tenth of that value annually—the result, according to analysts, of the EU's moratorium on the approval of new corn varieties already approved in the United States. Although one variety of biotech corn was approved by the EU prior to the moratorium, the United States grew other varieties. Thus, U.S. export of any corn to the EU was impractical because of the difficulty of segregating approved from unapproved varieties.

THE WTO CASE

The United States and co-complainants argued that the EU moratorium violated provisions of the WTO Agreement on the Application of Sanitary and Phytosanitary (SPS) Measures. The SPS agreement permits countries to regulate crops and food products to protect health and the environment, but their rules must be scientifically justified, and approval procedures must occur without undue delay. U.S. interests contend that there is no scientific evidence that GE food and feed crops are substantially different from, or any less safe than, conventional varieties, a conclusion they say even European scientific authorities have reached. The United States contends that EU biotechnology measures also are inconsistent with provisions of other WTO agreements, namely the Technical Barriers to Trade Agreement (TBT), the General Agreement on Tariffs and Trade (GATT 1994), and the Uruguay Round Agreement on Agriculture.

EU officials counter that their cautious approach to regulating biotechnology is necessary to cultivate trust among European consumers. At the same time, they also assert that they have shown good faith in moving to restart the approval process.

THE DISPUTE PANEL'S RULING

The WTO dispute panel found that the EU had maintained a moratorium on approvals of genetically engineered crops between June 1999 and the beginning of panel deliberations in August 2003.[5] The EU had claimed that there were no measures imposing a moratorium. The panel agreed with the United States and the co-complainants that marketing or import prohibitions on GE products in six EU countries (Austria, France, Germany, Greece, Italy, and Luxembourg) violated the WTO's Agreement on Sanitary and Phytosanitary (SPS) Measures, which requires such measures to be based on science and risk assessment. The panel upheld the complainants' charge that the EU also violated the SPS agreement by not ensuring that approvals for 24 of 27 genetically modified (GM) products in the approval pipeline were carried out without "undue delay."

The dispute panel's ruling dismissed several other U.S. and co-complainant claims, including claims that EU approval procedures were not based on appropriate risk assessment; that the EU unfairly applied different risk assessment standards for GE processing agents; and that the EU had unjustifiably discriminated between WTO members. The panel made no recommendations as to how the EU should bring its practices in line with WTO rules, nor did the ruling require the EU to change its regulatory framework for approving GE products. The EU announced on December 19, 2006, that it would not appeal the panel's ruling in the case. The United States and EU agreed that a reasonable period of time would be until November 21, 2007, a deadline subsequently extended by mutual agreement until January 11, 2008. Time would likely be required to reach intra-EU agreement to lift the prohibitions on already approved GE products or to launch court challenges to require member countries to lift their prohibitions on GE products.[6]

In May 2004, the EU effectively ended the moratorium by approving a GE corn variety (Syngenta Bt-11) for human consumption. Since then, the EU has approved GE corn varieties (Monsanto's NK603) for both human and animal consumption, Pioneer's 1570 corn for feed use, and 17 strains of GE corn seed (all derived from the MON810 strain approved in 1998) for commercial use.[7] The latest EU approval is the Amflora starch potato, approved for cultivation and for use in the production of industrial starch.[8] The Amflora potato is only the second GE variety approved for cultivation in 12 years; in 1998, the EU had approved MON810, a GE corn, for cultivation and for use as food and livestock feed.

While the United States and the EU are continuing technical discussions on market access issues for biotech products, both Canada and Argentina have settled their disputes with the EU. On July 15, 2009, Canada and the EU signed a final settlement of the WTO dispute that Canada had brought against the EU in May 2003. Similarly, Argentina and the EU announced their final settlement of the biotech dispute on March 18, 2010. All three parties have notified these settlements to the DSB as mutually agreed solutions. Both settlements provide for bilateral, biannual meetings between competent services of the European Commission and the co-complainants' authorities regarding the application of biotechnology to agriculture and related trade issues of mutual interest, including

- follow-up of authorizations of genetically modified products of interest to each of the parties;
- measures related to biotechnology that may affect trade between the parties, including measures adopted by EU member states;
- specific issues that arise in the context of requests for authorization submitted to regulatory evaluations;
- any trade impacts of asynchronous authorizations of genetically modified products;
- renewal of authorizations of genetically modified products; and
- exchanges of information regarding such issues as new legislation affecting biotechnological agriculture, or coordination mechanisms to solve eventual cases of adventitious presence of non-authorized GMOs in shipments of authorized products.

Differing Regulatory Approaches

The United States has embraced the concept of substantial equivalence with regard to GE products: if a GE product is substantially the same as its conventional counterpart, it should be regulated no differently than the conventional product. The EU, on the other hand, takes a "precautionary approach," which means that if scientific evidence is insufficient or inconclusive regarding a practice's or product's potential dangers to human or environmental health, it should be more vigorously regulated or even prohibited if there are reasonable grounds for concern, thus providing a safeguard against future unforeseen problems. Under this approach, the products of biotechnology are deemed to be inherently different from their conventional counterparts and to require a separate regulatory regime.

United States

The basic federal guidance for regulating biotechnology products is the *Coordinated Framework for Regulation of Biotechnology*.[9] One of its key principles is that GE products should be regulated according to their characteristics and unique features—not according to their method of production. Once approved, food products do not have to be labeled as to whether or not they contain any genetically modified organisms, except to the extent that a GE food is substantially different (e.g., contains an allergen or has a changed nutritional content). Because they are deemed substantially equivalent, GE products are regulated under existing federal statutory authorities.[10]

European Union

The EU has established separate structures specifically for approving GE products and for labeling products derived from them. Currently, the key measure is Council Directive 2001/18 (as amended in July 2003) which spells out steps for assessing human health and environmental risks before any GE product can be released into the environment or marketed.[11] Prior to the 2003 amendments, the competent authority in the EU member state where the product was to be released was responsible for assessing its safety and, if approved, notifying other member states, opening the way for marketing throughout the EU (with EU-level intervention if one member state disagreed with another's decision). The amended directive provides for a "one-door-one-key" approach, whereby the European Food Safety Authority (EFSA) conducts all scientific risk assessments and communicates risks to the public. Then, the EU Council of Ministers decides whether or not to approve a GE product for the EU market. EU regulations empower the European Commission to approve applications by default if the Council of Ministers fails to act on them within three months.

Labeling and Traceability

The WTO case does not involve EU labeling and traceability regulations, which, U.S. agricultural interests argue, continued to discriminate against U.S. exports even after the GE product approval moratorium was lifted. The labeling and traceability regulations, adopted in July 2003, require that most foods, ingredients, and (for the first time) animal feeds from GE products be labeled, even if they no longer contain detectable traces. The regulations (1830/2003 on the Traceability and Labeling of GMOs and 1829/2003 on Genetically Modified (GM) Food and Feed) were implemented in April 2004.

Under the regulations, a tolerance level for non-GE foods, feeds, and processed products of 0.9% is set for allowable "adventitious presence" (AP)—that is, unintended, low-level presence—of an EU-approved GE substance. All products with more than 0.9% must be labeled as containing GE products. Products like meat, milk, and eggs from animals fed or treated with GE materials will not have to be labeled, however. Traceability provisions now require all firms that produce, store, move, or process GE products to track and keep records on them from farm to consumer. Compliance with the EU labeling rule requires segregation of GE crops and foods derived from them from the time they are planted all the way through the processing and marketing chain. This entails prevention of pollen drift from GE to non-GE fields; and difficult and costly handling procedures such as using separate equipment, storage, and shipping containers, or at least painstakingly cleaning them. U.S. interests argue that food companies forced to label accurately all GE products face huge risks and liabilities. All of these problems discriminate against U.S. shipments—even though they are as safe as "conventional" shipments, they contend. In practice, many U.S. manufacturers have opted not to market GE products in the EU, in part due to the EU's stricter GE regulations.[12]

The European Commission is currently discussing new technical guidelines for the allowable level of genetically engineered organisms in food and feed shipments for GE products that have been approved in exporting countries but not the EU.

DIFFERING PUBLIC ATTITUDES?

Differing U.S. and EU perspectives may reflect the fact that U.S. consumers apparently have been not only less fearful of GE foods than their European counterparts, but also more confident in their food safety regulators. According to USDA's Economic Research Service (ERS), surveys of consumer attitudes toward GE products, conducted both here and overseas, have yielded mixed results. Still, "U.S. consumers have voiced little objection to genetically modified foods, while EU consumers have been vocal in their disapproval," ERS observed.[13]

Europeans may be much more wary of changes in how their food is produced due to a series of recent food safety crises during the 1990s. Bovine spongiform encephalopathy (BSE, or "mad cow disease") emerged in the United Kingdom and spread to other parts of Europe. U.K. food safety authorities first insisted that the disease could not be transmitted to humans eating meat from BSE-infected animals. By 1996 scientific evidence indicated there was a link between some cases of a similar human disease and consumption of BSE-contaminated beef. In 1999, high levels of dioxin were found in meat products and eggs originating in Belgium. Also, foot-and-mouth disease (FMD) outbreaks in Europe added to consumer concerns and to their "waning faith in regulatory agencies," according to the Pew Charitable Trust's Initiative on Food and Biotechnology. Pew is careful to point out that these crises have not been caused by GE food, but that GE food has been caught up in the general suspicion about food safety. Environmental groups in the EU, such as Greenpeace and Friends of the Earth, also have raised concerns about environmental impacts.

Recently, concerns have been voiced by European grain and feedstuffs traders and compound feed manufacturers about EU regulations and attitudes about GE corn and soybeans.[14] They point to short supplies and high prices for feedstuffs and call for changes in the EU regulations that impose strict limits on adventitious presence and that have helped create a shortage of feed supplies in the EU. COCERAL, the association of EU feedstuffs traders, has pointed out that neither Argentina nor Brazil, who along with the United States supply the bulk of annual EU feed supplies, could guarantee that its corn and soybean shipments contain only EU-approved GMOs.

CONGRESSIONAL INTEREST

In the last Congress, Members representing agricultural interests urged the United States to take an aggressive stance with respect to EU biotechnology regulations. Senator Grassley, the Ranking Member of the Senate Finance Committee, expressed hope that the EU would soon come into compliance with the WTO ruling in the biotechnology case. He noted that, although he would have preferred the U.S. Trade Representative to take a harder line in the case, he recognized that a "hard approach" of seeking retaliation might not be successful.[15] At the same time, many lawmakers are well aware of the risks involved in escalating U.S.-EU trade tensions to new heights.

How to proceed with the EU on biotechnology trade is an issue that the Bush Administration left unresolved. The options for the Obama Administration appear to be either to continue a dialogue with the EU on biotechnology policy and trade or to pursue retaliation for failure to comply with the earlier adverse ruling. Before the United States could impose retaliatory measures, however, it would have to request establishment of a panel to determine whether the EU had complied with the November 21, 2006, decision in the dispute. As the 111th Congress monitors the Administration's conduct of agricultural trade policy, the issues raised by the U.S.-EU biotechnology dispute will likely remain on the agenda.

APPENDIX. CHRONOLOGY OF THE U.S.-EU BIOTECH DISPUTE (DS291)

May 13, 2003	The United States requested consultation with the European Union concerning measures taken by EU and its member states affecting imports of biotech products from the United States.
August 7, 2003	The United States requested establishment of a panel.
August 18, 2003	The Dispute Settlement Body (DSB) deferred the establishment of a panel.
August 29, 2003	The DSB established a single panel to examine the U.S. dispute and similar disputes filed by Canada (DS292) and Argentina (DS293).
February 23, 2004	The United States, Canada and Argentina jointly requested the WTO Director- General to compose the panel.
March 4, 2004	The WTO Director-General composed the panel.
July 12, 2004	The chairman of the panel announced that it would not be able to complete its work in six months time, owing in part to the parties' request for additional time to prepare their rebuttals.
August 18, 2004	The chairman of the panel estimated that it would issue its final report by the end of March 2005, owing in part to the panel's decision to seek scientific and technical expert advice pursuant to Article 11 of the SPS Agreement and Article 13 of the Dispute Settlement Understanding (DSU).
November 2, 2005	The chairman of the panel informed the DSB that the panel would submit its final report to the parties by the end of June 2005, owing in part to a joint request of all parties that they be granted additional time to prepare further submissions to the panel,.
June 13, 2005	The chairman of the panel estimated that it would issue its final report to the parties by the end of October 2005.

August 11, 2005	The chairman of the panel estimated that it would issue its final report to the parties by the end of December 2005.
December 21, 2005	The chairman of the panel announced that it would require additional time to finish its report and estimated that it would issue its final report to the parties by the end of March 2006.
March 30, 2006	The chairman of the panel announced that it would not be possible to complete its report to the parties because the panel had not received comments from the parties on the interim final report. The panel estimated that it would issue its final report to the parties by mid-May 2006 and the final report would be circulated to members no later than the end of September 2006.
September 29, 2006	The panel reports were circulated to members.
November 21, 2006	The DSB adopted the panel report.
December 12, 2006	The EU announced its intention to implement the recommendations and rulings of the DSB in a manner consistent with its WTO obligations.
June 21, 2007	The United States and the EU notified the DSB that they had agreed that the reasonable period of time for the EU to implement the recommendations and rulings of the DSB would be 12 months (by November 21, 2007).
November 21, 2007	The parties informed the DSB that they had agreed to modify the reasonable period of time so as to expire on January 11, 2008.
January 14, 2008	The EU and the United States informed the DSB that they had reached an agreement on procedures under Article 21 and 22 of the DSU.
January 17, 2008	The United States requested authorization from the DSB to suspend concessions and other obligations.
February 6, 2008	The EU objected to the U.S. request for authorizations to suspend concessions and referred the matter to arbitration under Article 22.6 of the DSU.
February 15, 2008	The United States and the EU requested the arbitrator to suspend work pursuant to their agreed procedures under Article 21 and 22 of the DSU. The arbitrator suspended proceedings from February 18, 2008, until the United States requests their resumption under the circumstances agreed between the parties on January 14, 2008.
July 15, 2009	Canada and the European Union notified the DSB of a mutually agreed solution under Article 3.6 of the DSU. The parties agreed to establish a bilateral dialogue on agricultural biotech market access issues of mutual interest.
March 18, 2010	Argentina and the European Union notified the DSB of a mutually agreed solution under Article 3.6 of the DSU. As in the Canada-EU agreement, the parties agreed to establish a bilateral dialogue on agricultural biotech products of mutual interest.

End Notes

[1] The EU's labeling and traceability regulation is at http://europa.eu.int/eur-lex/pri/en/oj/dat/2003/l_268/ l_268 20031018 en00240028.pdf.

[2] Data are from ISAAA Brief 41-2009, *Global Status of Commercialized Biotech/GM Crops: 2009,* http://www.isaaa.org/resources/publications/briefs/41/executivesummary/default.asp.

[3] USDA, Economic Research Service website, *Adoption of Genetically Engineered Crops in the U.S.*, at http://www.ers.usda.gov/data/BiotechCrops/.

[4] Before the moratorium, the EU had approved the commercial release of 18 genetically modified organisms (GMOs), including MON810 and "Roundup-Ready" soybeans. Until the EU's approval in March 2010 of a potato variety for cultivation for industrial use, the corn variety MON810 was the only GE product approved for cultivation in the EU.

[5] Background on the U.S.-EU biotechnology dispute is available at http://www.wto.org/english/tratop_e/dispu_e/ cases_e/ds291_e.htm.

[6] See "EU Council Rejects Proposals to Lift Bans on GMO Products" in *Inside U.S. Trade*, December 22, 2006, at http://www.insidetrade.com/secure/dsply_nl_txt.asp?f=wto2002.ask&dh=70507021&q=gmo.

[7] For a list of GE products authorized or pending authorization in the EU, see http://www.gmo-compass .org/eng/gmo/ db/.

[8] BASF Plant Science, News Release, "European Commission Approves Amflora Starch Potato," March 2, 2010, P179/10, at http://www.basf.com/group/corporate/en/function/conversions:/publish/content/news-and-media-relations/ news-releases/downloads/2010/P179-Amflora-e.pdf.

[9] Executive Office of the President, Office of Science and Technology Policy (51 FR 23302), *Coordinated Framework for Regulation of Biotechnology*, June 26, 1986, viewed at http://usbiotechreg.nbii.gov/ CoordinatedFrameworkForRegulationOfBiotechnology1986.pdf; see also CRS Report RL32809, *Agricultural Biotechnology: Background and Recent Issues*, by Tadlock Cowan and Geoffrey S. Becker.

[10] See http://heinonline.org/HOL/Page?handle=hein.fedreg/051123&id=96&collection=fedreg.

[11] See the EU's *Directive on the release of genetically modified organisms*, together with amending acts, at http://europa.eu/scadplus/leg/en/lvb/l28130.htm.

[12] Personal communication with Economic Research Service, January 21, 2005.

[13] Details on these surveys can be found at ERS, *Economic Issues in Agricultural Biotechnology* (Information Bulletin No. 762), February 2001, pp. 28-30. In December 2006, the Pew Initiative on Food and Biotechnology released a new survey of U.S. attitudes on GM foods which, Pew said, determined that Americans' knowledge of them remains low, that their opposition to such foods has softened somewhat since 2001 (from 58% to 48%), and support remains stable at 27%. Survey results can be found at http://pewagbiotech.org/research/2006update/2.php.

[14] "GMO zero tolerance devastating for EU feed industry," from *AgBioWorld*, July 19, 2007, viewed at http:// www.agbioworld.org/newsletter_wm/index.php?caseid=archive&newsid=2747.

[15] "U.S. To Hold Off on GMO Retaliation Against EU, France Bans GMO Corn", *Inside U.S. Trade*, January 18, 2008, viewed at http://www.insidetrade.com/secure/dsply_nl_txt.asp?f=wto2002.ask&dh=107026683&q=.

In: Technology Policies and Issues for the U.S. Government ISBN: 978-1-61209-501-1
Editors: Danielle A. Godden and Mason C. Andrews

Chapter 6

INDUSTRIAL COMPETITIVENESS AND TECHNOLOGICAL ADVANCEMENT: DEBATE OVER GOVERNMENT POLICY *

Wendy H. Schacht

SUMMARY

There is ongoing interest in the pace of U.S. technological advancement due to its influence on U.S. economic growth, productivity, and international competitiveness. Because technology can contribute to economic growth and productivity increases, congressional attention has focused on how to augment private-sector technological development. Legislative activity over the past 25 or more years has created a policy for technology development, albeit an ad hoc one. Because of the lack of consensus on the scope and direction of a national policy, Congress has taken an incremental approach aimed at creating new mechanisms to facilitate technological advancement in particular areas and making changes and improvements as necessary.

Congressional action has mandated specific technology development programs and obligations in federal agencies. Many programs were created based upon what individual committees judged appropriate within the agencies over which they had authorization or appropriation responsibilities. However, there has been recent legislative activity directed at eliminating or significantly curtailing many of these federal efforts. Although, for the most part, this approach has not been adopted, the budgets for several programs have declined.

The proper role of the federal government in technology development and the competitiveness of U.S. industry continues to be a topic of congressional debate. Current legislation affecting the R&D environment have included both direct and indirect measures to facilitate technological innovation. In general, direct measures are those which involve budget

* This is an edited, reformatted and augmented version of Congressional Research Service publication, Report RL33528, dated February 23, 2010.

outlays and the provision of services by government agencies. Indirect measures include financial incentives and legal changes (e.g., liability or regulatory reform; new antitrust arrangements). As the 111th Congress develops its budget priorities, the manner by which the government encourages technological progress in the private sector again may be explored and/or redefined.

Technology and Competitiveness

Technological advancement in U.S. industry often has been supported by congressional initiatives over the past 25 or more years. This approach has involved both direct measures that concern budget outlays and the provision of services by government agencies (such as the now terminated Advanced Technology Program (ATP) and the Manufacturing Extension Partnership (MEP) of the National Institute of Standards and Technology) and indirect measures that include financial incentives and legal changes. Many of these efforts, however, have been revisited over the past several congresses. Indirect strategies such as tax policies, intellectual property right protection, and antitrust laws to promote technological advancement and increased government support for basic research appear to have been favored over direct federal funding for private sector technology commercialization initiatives.

Interest in technology development and industrial innovation increased as concern mounted over the economic strength of the nation and over competition from abroad. For the United States to be competitive in the world economy, U.S. companies must be able to engage in trade, retain market shares, and offer high quality products, processes, and services while the nation maintains economic growth and a high standard of living. Technological advancement is important because the commercialization of inventions provides economic benefits from the sale of new products or services; from new ways to provide a service; or from new processes that increase productivity and efficiency. It is widely accepted that technological progress is responsible for up to one-half the growth of the U.S. economy, and is one principal driving force in long-term growth and increases in living standards.

Technological advances can further economic growth because they contribute to the creation of new goods, new services, new jobs, and new capital. The application of technology can improve productivity and the quality of products. It can expand the range of services that can be offered as well as extend the geographic distribution of these services. The development and use of technology also plays a major role in determining patterns of international trade by affecting the comparative advantages of industrial sectors. Since technological progress is not necessarily determined by economic conditions—it also can be influenced by advances in science, the organization and management of firms, government activity, or serendipity—it can have effects on trade independent of shifts in macroeconomic factors. New technologies also can help compensate for possible disadvantages in the cost of capital and labor faced by firms.

THE FEDERAL ROLE IN TECHNOLOGY DEVELOPMENT

In the recent past, American companies faced increased competitive pressures in the international marketplace from firms based in countries where governments actively promote commercial technological development and application. In the United States, the generation of technology for the commercial marketplace is primarily a private sector activity. The federal government traditionally becomes involved only for certain limited purposes. Typically these are activities which have been determined to be necessary for the "national good" but which cannot, or will not, be supported by industry.

To date, the U.S. government has funded research and development (R&D) to meet the mission requirements of the federal departments and agencies. It also finances efforts in areas where there is an identified need for research, primarily basic research, not being performed in the private sector. Federal support reflects a consensus that basic research is critical because it is the foundation for many new innovations. However, any returns created by this activity are generally long term, sometimes not marketable, and not always evident. Yet the rate of return to society as a whole generated by investments in research is significantly larger than the benefits that can be captured by the firm doing the work.[1]

Many past government activities to increase basic research were based on a "linear" model of innovation. This theory viewed technological advancement as a series of sequential steps starting with idea origination and moving through basic research, applied research, development, commercialization, and diffusion into the economy. Increases in federal funds in the basic research stage were expected to result in concomitant increases in new products and processes. However, this linear concept is no longer considered valid. Innovations often occur that do not require basic or applied research or development; in fact most innovations are incremental improvements to existing products or processes. In certain areas, such as biotechnology, the distinctions between basic research and commercialization are small and shrinking. In others, the differentiation between basic and applied research is artificial. The critical factor is the *commercialization* of the technology. Economic benefits accrue only when a technology or technique is brought to the marketplace where it can be sold to generate income or applied to increase productivity. Yet, while the United States has a strong basic research enterprise, foreign firms appear equally, if not more, adept at taking the results of these scientific efforts and making commercially viable products. Often U.S. companies are competing in the global marketplace against goods and services developed by foreign industries from research performed in the United States. Thus, there has been increased congressional interest in mechanisms to accelerate the development and commercialization processes in the private sector.

The development of a governmental effort to facilitate technological advance has been particularly difficult because of the absence of a consensus on the need for an articulated policy. Technology demonstration and commercialization have traditionally been considered private sector functions in the United States. While over the years there have been various programs and policies to facilitate such activities(such as tax credits, technology transfer to industry, and patents), the approach had been ad hoc and uncoordinated. Much of the program development was based upon what individual committees judged appropriate for the agencies over which they have jurisdiction. Despite the importance of technology to the economy, technology-related considerations often have not been integrated into economic decisions.

There have been attempts to provide a central focus for governmental activity in technology matters. P.L. 100-519 created within the Department of Commerce a Technology Administration headed by a new Under Secretary for Technology. However, this office was abolished as of the end of FY2007. In November 1993, former President Clinton established a National Science and Technology Council to coordinate and integrate decisionmaking in science and technology at all policy levels. However, technological issues and responsibilities remain shared among many departments and agencies. This diffused focus has sometimes resulted in actions which, if not at cross purposes, may not have accounted for the impact of policies or practices in one area on other parts of the process. Technology issues involve components which operate both separately and in concert. While a diffused approach can offer varied responses to varied issues, the importance of interrelationships may be underestimated and their usefulness may suffer.

Several times, Congress has examined the idea of an industrial policy to develop a coordinated approach on issues of economic growth and industrial competitiveness. Technological advance is both one aspect of this and an altogether separate consideration. In looking at the development of an identified policy for industrial competitiveness, advocates argue that such an effort could ameliorate much of the uncertainty with which the private sector perceives future government actions. Some commentators have argued that consideration and delineation of national objectives could encourage industry to engage in more long-term planning with regard to R&D and to make decisions as to the best allocation of resources. Such a technology policy could generate greater consistency in government activities. Because technological development involves numerous risks, efforts to minimize uncertainty regarding federal programs and policies may help alleviate some of the disincentives perceived by industry.

The development of a technology policy, however, is a contentious issue. There is widespread resistance to what could be and has been called national planning, due variously to doubts as to its efficacy, to fear of adverse effects on our market system, to political beliefs about government intervention in our economic system, and to the current emphasis on short-term returns in both the political and economic arenas. Opponents of a national industrial policy may see this approach as government interference in the marketplace to "pick winners and losers." Instead, it is argued, measures that would occasion a better investment environment for industry to expand innovation-related efforts would be preferable to government decisionmaking in technological advancement.

Consideration of what constitutes government policy (both in terms of the industrial policy and technology policy) covers a broad range of ideas from laissez-faire to special government incentives to target specific high-technology, high-growth industries. Suggestions have been made for the creation of federal mechanisms to identify and support strategic industries and technologies. Various federal agencies and private sector groups have developed critical technology lists. However, others maintain that such targeting is an unwanted, and unwarranted, interference in the private sector which will cause unnecessary dislocations in the marketplace or a misallocation of resources. From their perspective, the government does not have the knowledge or expertise to make business-related decisions. Instead, they argue, the appropriate role for government is to encourage innovative activities in all industries and to keep market related decisionmaking within the business community that has ultimate responsibility for commercialization and where such decisions have traditionally been made.

The relationship between government and industry often is a major factor affecting innovation and the environment within which technological development takes place. This relationship can be adversarial, with the government acting to regulate or restrain the business community, rather than to facilitate its positive contributions to the nation. However, this may be changing as the benefits of industry/government cooperation become more apparent. There are an increasing number of areas where the traditional distinctions between public and private sector functions and responsibilities are becoming blurred. Many assumptions have been questioned, particularly in light of the increased internationalization of the U.S. economy. The business sector is no longer viewed in an exclusively domestic context; the economy of the United States is often tied to the economies of other nations. The technological superiority long held by the United States in many areas has been challenged by other industrialized countries in which economic, social, and political policies and practices foster government-industry cooperation in technological development.

The approach taken by the former Clinton Administration was a divergence from the past. Articulated in two reports issued in February 1993 (*A Vision of Change for America and Technology for America's Economic Growth, A New Direction to Build Economic Strength*),[2] the proposal called for a national commitment to, and a strategy for, technological advancement as part of a defined national economic policy. This detailed strategy offered a policy agenda for economic growth in the United States, of which technological development and industrial competitiveness were critical components.

In articulating a national technology policy, the approach initially recommended and subsequently followed by the Clinton Administration provided a wide range of options while for the most part reflecting then current trends in congressional efforts to facilitate industrial advancement. This policy, backed by congressional legislation, increased federal coordination and augmented direct government spending for technological development. While many past activities focused primarily on research, the new initiatives shifted the emphasis toward *development* of new products, processes, and services by the private sector for the commercial marketplace. In addition, a significant number of the proposals aimed to increase both government and private sector support for R&D leading to the commercialization of technology.

This approach was questioned by recent congresses and by the Bush Administration. Instead, policies appeared more supportive of indirect strategies such as tax incentives, intellectual property protection, and antitrust laws to promote technology advancement, increased government support for basic research, and decreased direct federal funding for private sector technology activities. In the 2006 State of the Union Address, former President Bush announced the "American Competitiveness Initiative" to facilitate innovation and provide "our nation's children a firm grounding in math and science." To achieve these goals, the President called for doubling over the next 10 years the amount of federal funding for basic research, particularly in the National Science Foundation, the Office of Science in the Department of Energy, and in the core programs of the National Institute of Standards and Technology, Department of Commerce. In addition, the Initiative would increase the number of math and science teachers and make permanent the research and experiment tax credit.

Despite the continuing debate on what is the appropriate role of government and what constitutes a suitable government technology development policy, it remains an undisputed fact that what the government does or does not do affects the private sector and the marketplace. The various rules, regulations, and other activities of the government have become de facto policy as they relate to, and affect, innovation and technological advancement.

Legislative Initiatives and Current Programs

Legislative initiatives have reflected a trend toward expanding the government's role beyond traditional funding of mission-oriented R&D and basic research toward the facilitation of technological advancement to meet other critical national needs, including the economic growth that flows from new commercialization and use of technologies and techniques in the private sector. An overview of recent legislation shows federal efforts aimed at (1) encouraging industry to spend more on R&D; (2) assisting small high-technology businesses; (3) promoting joint research activities between companies; (4) fostering cooperative work between industry and universities; (5) facilitating the transfer of technology from the federal laboratories to the private sector; and (6) providing incentives for quality improvements. These efforts tend toward removing barriers to technology development in the private sector (thereby permitting market forces to operate) and providing incentives to encourage increased private sector R&D activities. While most focus primarily on research, some also involve policies and programs associated with technology development and commercialization.

Increased R&D Spending

To foster research in the private sector, Congress created a temporary tax credit for incremental increases in qualified research spending. The 1981 Economic Recovery Tax Act (P.L. 97-34) provided a 25% tax credit for the increase in a firm's qualified research costs above the average expenditures for the previous three tax years. Qualified costs included in-house expenditures such as wages for researchers, material costs, and payments for use of equipment; 65% of corporate grants towards basic research at universities and other relevant institutions; and 65% of payments for contract research. The credit applied to research expenditures through 1985. While never made permanent, the Research and Experimentation Tax Credit has been extended many times. Several changes have been made to the rate and to the definition of qualified expenses. Most recently, P.L. 110-343 retroactively extended the credit through the end of calendar year 2009.[3]

In 1982, the Small Business Development Act (P.L. 97-219) established Small Business Innovation Research (SBIR) programs within the major federal R&D agencies designed to increase participation of small, innovative companies in federally funded research and development.[4] Extended several times, the program requires that a set percentage of each agency's applicable extramural R&D budget—originally 1.23%, now 2.5%—is to be used to support mission-related work in small companies. Funding is, in part, dependent on companies obtaining private sector support for the commercialization of the resulting products or processes. The authorization for the program terminated on September 30, 2008 but has been temporarily extended five times and is now operating through April 30, 2010. In 1992, a pilot effort, the Small Business Technology Transfer (STTR) program, was created to encourage firms to work with universities or federal laboratories to commercialize the results of research. Initially funded by a 0.15% (phased in) set-aside, the STTR program has been reauthorized several times and the set-aside raised to 0.3% in FY2004. This program sunset at the end of FY2009, but has been temporarily extended through April 30, 2010.

The Omnibus Trade and Competitiveness Act of 1988 (P.L. 100-418) created the Advanced Technology Program (ATP) at the Department of Commerce's National Institute of Standards and Technology (NIST).[5] ATP provided seed funding, matched by private sector investment, for companies or consortia of universities, industries, and/or government laboratories to accelerate development of generic technologies with broad application across industries. The first awards were made in 1991. As of the end of 2007, when ATP was terminated and replaced by the Technology Innovation Program, 824 projects had been funded representing approximately $1.6 billion in federal dollars matched by $1.5 billion in private sector financing. About 68% of the awardees were small businesses or cooperative efforts led by such firms; 227 projects were joint ventures.[6]

The Technology Innovation Program was created by P.L. 110-69, the America COMPETES Act.[7] While similar to ATP in the intent to promote high-risk R&D that would be of broad-based economic benefit to the nation, there are several differences in the operation of the new activity. Funding under TIP is limited to small and medium-sized businesses whereas grants under ATP were available to companies regardless of size. In addition, in the Advanced Technology Program, joint ventures were required to include two separately owned for-profit firms and could include universities, government laboratories, and other research establishments as participants in the project but not as recipients of the grant. In the TIP initiative, a joint venture may involve two separately owned for-profit companies but may also be comprised of one small or medium-sized firm and a university (or other non-profit research organization). A single company could receive up to $2 million for up to three years under ATP; under TIP, the participating company (which must be a small or medium-sized business) may receive up to $3 million for up to three years. In ATP, small and medium-sized companies were not required to cost share (large firms provided 60% of the total cost of the project), while in TIP there is a 50% cost sharing requirement which, again, only applies to the small and medium-sized businesses that are eligible. There were no funding limits for the five-year funding available for joint ventures under ATP; the TIP limits joint venture funding to $9 million for up to five years. The Advisory Board that was created to assist in the Advanced Technology Program included industry representatives as well as federal government personnel and representatives from other research organizations. The Advisory Board for the Technology Innovation Program is to be comprised of only private sector members.

In January 2009, nine awards were announced for "new research projects to develop advanced sensing technologies that would enable timely and detailed monitoring and inspection of the structural health of bridges, roadways and water systems that comprise a significant component of the nation's public infrastructure." According to TIP, $42.5 million in federal money is expected to be matched by $45.7 in private sector support. Twenty additional awards were announced in December 2009 totaling almost $71.0 million in NIST financing with approximately $145.7 million in funding from other sources.

Industry-University Cooperative Efforts

The promotion of cooperative efforts among academia and industry is aimed at increasing the potential for the commercialization of technology.[8] Traditionally, basic research has been performed in universities or in the federal laboratory system while the business community focuses on the manufacture or provision of products, processes, or services. Universities are especially suited to undertake basic research. Their mission is to educate and basic research is an integral part of the educational process. Universities generally are able to undertake these activities because they do not have to produce goods for the marketplace and therefore can do research not necessarily tied to the development of a commercial product or process.

Subsequent to World War II, the federal government supplanted industry as the primary source of funding for basic research in universities. It also became the principal determinant of the type and direction of the research performed in academia. This resulted in a disconnect between the university and industrial communities. The separation and isolation of the parties involved in the innovation process is thought by many observers to be a barrier to technological progress. The difficulties in moving an idea from the concept stage to a commercial product or process may be compounded when several entities are involved. Legislation to stimulate cooperative efforts among those involved in technology development has been viewed as one way to promote innovation and facilitate the international competitiveness of U.S. industry.

Several laws have attempted to encourage industry-university cooperation. Title II of the Economic Recovery Tax Act of 1981 (P.L. 97-34) provided, in part, a 25% tax credit for 65% of all company payments to universities for the performance of basic research.[9] Firms were also permitted a larger tax deduction for charitable contributions of equipment used in scientific research at academic institutions. The Tax Reform Act of 1986 (P.L. 99-514) kept this latter provision, but reduced the credit for university basic research to 20% of all corporate expenditures for this over the sum of a fixed research floor plus any decrease in non-research giving.

The 1981 act also provided an increased charitable deduction for donations of new equipment by a manufacturer to an institution of higher education. This equipment must be used for research or research training for physical or biological sciences within the United States. The tax deduction is equal to the manufacturer's cost plus one-half the difference between the manufacturer's cost and the market value, as long as it does not exceed twice the cost basis. Extended numerous times, with certain changes, the research and experimentation tax credit is due to expire at the end of calendar year 2009.

An additional legislative initiative to foster interaction between academia and the business community is contained in amendments to the patent and trademark laws , commonly referred to as the "Bayh-Dole Act."[10] This law is intended to use patent ownership as an incentive for private sector development and commercialization of federally supported R&D. Title to inventions made by contractors receiving federal research funds is to be vested in the contractor if they are small businesses, universities, or not-for-profit institutions. Certain rights to the patent are reserved for the government and these organizations are required to commercialize within a predetermined and agreed upon time frame. Providing universities with patent title is expected to encourage licensing to industry where the technology can be manufactured or used thereby creating a financial return to the academic institution. University patent applications and licensing have increased significantly since this law was enacted.

Subsequently, the CREATE Act, P.L. 108-453, made changes in the patent laws to further promote cooperative research and development among universities, government, and the private sector. The bill amends section 103(c) of title 25, United States Code, such that certain actions between researchers under a joint research agreement will not preclude patentability.[11]

Joint Industrial Research

Private sector investments in basic research are often costly, long term, and risky. Although not all advances in technology are the result of research, it is often the foundation of important new innovations. To encourage increased industrial involvement in research, legislation was enacted to allow for joint ventures in this arena. It is argued that cooperative research reduces risks and costs and allows for work to be performed that crosses traditional boundaries or expertise and experience. Such collaborative efforts make use of existing and support the development of new resources, facilities, knowledge, and skills.

The National Cooperative Research Act (P.L. 98-462) encourages companies to undertake joint research. The legislation clarifies the antitrust laws and requires that a "rule of reason" standard be applied in determinations of violations of these laws; cooperative research ventures are not to be judged illegal "per se." It eliminates treble damage awards for those research ventures found in violation of the antitrust laws if prior disclosure (as defined in the law) has been made. P.L. 98-462 also makes changes in the way attorney fees are awarded. Defendants can collect attorney fees in specified circumstances, including when the claim is judged frivolous, unreasonable, without foundation, or made in bad faith. However, the attorney fee award to the prevailing party may be offset if the court decides that the prevailing party conducted a portion of the litigation in a manner which was frivolous, unreasonable, without foundation, or in bad faith. These provisions were included to discourage frivolous litigation against joint research ventures without simultaneously discouraging suits of plaintiffs with valid claims. Between 1985 (when the law went into effect) and 2003, 913 joint research ventures have filed with the Department of Justice.

P.L. 103-42, the National Cooperative Production Amendments Act of 1993, amended the National Cooperative Research Act by, among other things, extending the original law's provisions to joint manufacturing ventures. These provisions are only applicable, however, to cooperative production when (1) the principal manufacturing facilities are "located in the United States or its territories, and (2) each person who controls any party to such venture ... is a United States person, or a foreign person from a country whose law accords antitrust treatment no less favorable to United States persons than to such country's domestic persons with respect to participation in joint ventures for production."

Commercialization of the Results of Federally Funded R&D

Another approach to encouraging the commercialization of technology involves the transfer of technology from federal laboratories and contractors to the private sector where commercialization can proceed. Because the federal laboratory system has extensive science and technology resources and expertise developed in pursuit of mission responsibilities, it is a potential source of new ideas and knowledge which may be used in the business community.[12]

Despite the potential offered by the resources of the federal laboratory system, however, the commercialization level of the results of federally funded R&D remained low. Studies indicated that only approximately 10% of federally owned patents were ever utilized. There are many reasons for this low level of usage, one of which is the fact that some technologies and/or patents have no market application. However, industry unfamiliarity with these technologies, the "not-invented-here" syndrome, and perhaps more significantly, the ambiguities associated with obtaining title to or exclusive license to federally owned patents also contribute to the low level of commercialization.

Over the years, several governmental efforts have been undertaken to augment industry's awareness of federal R&D resources. The Federal Laboratory Consortium for Technology Transfer was created in 1972 (from a Department of Defense program) to assist in transferring technology from the federal government to state and local governments and the private sector. To expand on the work of the Federal Laboratory Consortium, and to provide added emphasis on the commercialization of government technology, Congress passed P.L. 96-480, the Stevenson-Wydler Technology Innovation Act of 1980. Prior to this law, technology transfer was not an explicit mandate of the federal departments and agencies with the exception of the National Aeronautics and Space Administration. To provide "legitimacy" to the numerous technology activities of the government, Congress, with strong bipartisan support, enacted P.L. 96-480 which explicitly states that the federal government has the responsibility, "to ensure the full use of the results of the nation's federal investment in research and development." Section 11 of the law created a system within the federal government to identify and disseminate information and expertise on what technologies or techniques are available for transfer. Offices of Research and Technology Applications were established in each federal laboratory to distinguish technologies and ideas with potential applications in other settings.

Several amendments to the Stevenson-Wydler Technology Innovation Act have been enacted to provide additional incentives for the commercialization of technology. P.L. 99-502, the Federal Technology Transfer Act, authorizes activities designed to encourage industry, universities, and federal laboratories to work cooperatively. It also establishes incentives for federal laboratory employees to promote the commercialization of the results of federally funded research and development. The law amends P.L. 96-480 to allow government-owned, government-operated laboratories to enter into cooperative R&D agreements (CRADAs) with universities and the private sector. This authority is extended to government-owned, contractor-operated laboratories by the Department of Defense FY1990 Authorization Act, P.L. 101-189.[13] Companies, regardless of size, are allowed to retain title to inventions resulting from research performed under cooperative agreements. The federal government retains a royalty-free license to use these patents. The Technology Transfer Improvements and Advancement Act (P.L. 104-113), clarifies the dispensation of intellectual property rights under CRADAs to facilitate the implementation of these cooperative efforts. The Federal Laboratory Consortium is given a legislative mandate to assist in the coordination of technology transfer. To further promote the use of the results of federal R&D, certain agencies are mandated to create a cash awards program and a royalty sharing activity for federal scientists, engineers, and technicians in recognition of efforts toward commercialization of this federally developed technology. These efforts are facilitated by a provision of the National Defense Authorization Act for FY1991 (P.L. 101-510), which amends the Stevenson-Wydler Technology Innovation Act to allow government agencies and laboratories to develop partnership intermediary programs to augment the transfer of laboratory technology to the small business sector.

Amendments to the Patent and Trademark law contained in Title V of P.L. 98-620 made changes which are designed to improve the transfer of technology from the federal laboratories—especially those operated by contractors—to the private sector and increase the chances of successful commercialization of these technologies. This law permits the contractor at government-owned, contractor-operated laboratories (GOCOs) to make decisions at the laboratory level as to the granting of licenses for subject inventions. This has the potential of effecting greater interaction between laboratories and industry in the transfer of technology. Royalties on these inventions are also permitted to go back to the laboratory contractor to be used for additional R&D, awards to individual laboratory inventors, or education. While there is a cap on the amount of the royalty returning directly to the lab in order not to disrupt the agency's mission requirements and congressionally mandated R&D agenda, the establishment of discretionary funds gives contractor-operated laboratories added incentive to encourage technology transfer.

Under P.L. 98-620, private companies, regardless of size, are allowed to obtain exclusive licenses for the life of the patent. Prior restrictions allowed large firms use of exclusive license for only 5 of the 17 years (now 20 years) of the life of the patent. This was expected to encourage improved technology transfer from the federal laboratories or the universities (in the case of university operated GOCOs) to large corporations which often have the resources necessary for development and commercialization activities. In addition, the law permits GOCOs (those operated by universities or nonprofit institutions) to retain title to inventions made in the laboratory within certain defined limitations. Those laboratories operated by large companies are not included in this provision.

P.L. 106-404, the Technology Transfer Commercialization Act, altered practices concerning patents held by the government to make it easier for federal agencies to license such inventions. The law amends the Stevenson-Wydler Technology Innovation Act and the Bayh-Dole Act to decrease the time delays associated with obtaining an exclusive or partially exclusive license. Previously, agencies were required to publicize the availability of technologies for three months using the *Federal Register* and then provide an additional 60 day notice of intent to license by an interested company. Under this legislation, the time period was shortened to 15 days in recognition of the ability of the internet to offer widespread notification and the necessity of time constraints faced by industry in commercialization activities. Certain rights are retained by the government. The bill also allows licenses for existing government-owned inventions to be included in CRADAs.

The program of regional centers to assist small manufacturing companies use knowledge and technologies developed under the auspices of the National Institute of Standards and Technology and other federal agencies, created by the Omnibus Trade and Competitiveness Act, was also intended to promote the commercialization of federally supported R&D.[14] Federal funding for the centers is matched by non-federal sources including state and local governments and industry. Originally, seven Regional Centers for the Transfer of Manufacturing Technology were selected. The initial program was expanded in 1994 to create the Manufacturing Extension Partnership (MEP) to meet new and growing needs of the community. In a more varied approach, the Partnership involves both large centers and smaller, more dispersed organizations sometimes affiliated with larger centers as well as the NIST State Technology Extension Program which provides states with grants to develop the infrastructure necessary to transfer technology from the federal government to the private sector and a program which electronically ties the disparate parties together along with other federal, state, local, and academic technology transfer organizations. There are now centers in all 50 states and Puerto Rico. Since the manufacturing extension activity was created in 1989, awards made by NIST have resulted in the creation of approximately 400 regional offices. [It should be noted that the Department of Defense also funded 36 centers through its Technology Reinvestment Project (TRP) in FY1994 and FY1995. When the TRP was terminated, NIST took over support for 20 of these programs in FY1996 and funded the remaining efforts during FY1997.]

A new NIST program of partnerships between industry and other educational or research institutions to develop new manufacturing processes, techniques, or materials was authorized, but not funded, by the American COMPETES Act. In addition, a manufacturing fellowship program would be created with stipends available for post-doctoral work at NIST. These activities would differ from the established MEP effort where no new manufacturing research is conducted as existing manufacturing technology is applied to the needs of small and medium-sized firms.

Different Approach?

As indicated above, the laws affecting the R&D environment have included both direct and indirect measures to facilitate technological innovation. In general, direct measures are those which involve budget outlays and the provision of services by government agencies. Indirect measures include financial incentives and legal changes (e.g., liability or regulatory reform; new antitrust arrangements). Supporters of indirect approaches argue that the market is superior to government in deciding which technologies are worthy of investment. Mechanisms that enhance the market's opportunities and abilities to make such choices are preferred. Advocates further state that dependency on agency discretion to assist one technology in preference to another will inevitably be subjected to political pressures from entrenched interests. Proponents of direct government assistance maintain, conversely, that indirect methods can be wasteful and ineffective and that they can compromise other goals of public policy in the hope of stimulating innovative performance. Advocates of direct approaches argue that it is important to put the country's scarce resources to work on those technologies that have the greatest promise as determined by industry and supported by its willingness to match federal funding.

In the past, participants in the debates generally did not make definite (or exclusionary) choices between the two approaches, nor consistently favor one over the other. For example, some proponents of a stronger direct role for the government in innovation also have supported enhanced tax preferences for R&D spending, an indirect mechanism. Opponents of direct federal funding for specific projects (e.g., SEMATECH, flat panel displays) may nevertheless back similar activities focused on more general areas such as manufacturing or information technology. However, in recent congresses, legislators directed many of their efforts toward eliminating or curtailing some of the programs that previously had enjoyed bipartisan support. Initiatives to terminate the Advanced Technology Program, funding for flat panel displays, and agricultural extension reflected concern about the role of government in developing commercial technologies. While funding for several programs decreased, and ATP was replaced by the Technology Innovation Program, support for most ongoing activities continued. How the debate over federal funding evolves in the 111th Congress may serve to redefine thinking about the government's efforts in promoting technological advancement in the private sector.

End Notes

[1] Edwin Mansfield, "Social Returns From R&D: Findings, Methods, and Limitations," *Research/Technology Management*, November-December 1991, 24. See also Charles I. Jones and John C. Williams, "Measuring the Social Return to R&D," *Quarterly Journal of Economics*, November 1998, 1119 and Richard R. Nelson and Paul M. Romer, "Science, Economic Growth, and Public Policy," in Bruce R. Smith and Claude E. Barfield, eds. *Technology, R&D, and the Economy*, (Washington, The Brookings Institution and the American Enterprise Institute, Washington, 1996), 57.

[2] Available from author.

[3] CRS Report RL31181, *Research and Experimentation Tax Credit: Current Status and Selected Issues for Congress*, by Gary Guenther

[4] See CRS Report 96-402, *Small Business Innovation Research (SBIR) Program*, by Wendy H. Schacht and CRS Report RS22865, *The Small Business Innovation Research (SBIR) Program: Reauthorization Efforts*, by Wendy H. Schacht

[5] See CRS Report 95-36, *The Advanced Technology Program*, by Wendy H. Schacht

[6] National Institute of Standards and Technology, Historical Statistics on ATP Awards/Winners, available at http://www.atp.nist.gov/eao/statisticshtm

[7] See CRS Report RS22815, *The Technology Innovation Program*, by Wendy H. Schacht

[8] For more information see CRS Report RL33526, *Cooperative R&D: Federal Efforts to Promote Industrial Competitiveness*, by Wendy H. Schacht

[9] See CRS Report RL31181, *Research and Experimentation Tax Credit: Current Status and Selected Issues for Congress*, by Gary Guenther

[10] See CRS Report RL32076, *The Bayh-Dole Act: Selected Issues in Patent Policy and the Commercialization of Technology*, by Wendy H. Schacht and CRS Report RL30320, *Patent Ownership and Federal Research and Development (R&D): A Discussion on the Bayh-Dole Act and the Stevenson-Wydler Act*, by Wendy H. Schacht

[11] See CRS Report RS21882, *Collaborative R&D and the Cooperative Research and Technology Enhancement (CREATE) Act*, by Wendy H. Schacht

[12] See CRS Report RL33527, *Technology Transfer: Use of Federally Funded Research and Development*, by Wendy H. Schacht

[13] See CRS Report 95-150, *Cooperative Research and Development Agreements (CRADAs)*, by Wendy H. Schacht

[14] See CRS Report 97-104, *Manufacturing Extension Partnership Program: An Overview*, by Wendy H. Schacht

In: Technology Policies and Issues for the U.S. Government ISBN: 978-1-61209-501-1
Editors: Danielle A. Godden and Mason C. Andrews

Chapter 7

THE FEDERAL NETWORKING AND INFORMATION TECHNOLOGY RESEARCH AND DEVELOPMENT PROGRAM: FUNDING ISSUES AND ACTIVITIES *

Patricia Moloney Figliola

SUMMARY

In the early 1990s, Congress recognized that several federal agencies had ongoing high-performance computing programs, but no central coordinating body existed to ensure long-term coordination and planning. To provide such a framework, Congress passed the High-Performance Computing and Communications Program Act of 1991 (P.L. 102-194) to enhance the effectiveness of the various programs. In conjunction with the passage of the act, the White House Office of Science and Technology Policy (OSTP) released *Grand Challenges: High-Performance Computing and Communications*. That document outlined a research and development (R&D) strategy for high-performance computing and a framework for a multiagency program, the High-Performance Computing and Communications (HPCC) Program. The HPCC Program has evolved over time and is now called the Networking and Information Technology Research and Development (NITRD) Program, to better reflect its expanded mission.

Proponents assert that federal support of information technology (IT) R&D has produced positive outcomes for the country and played a crucial role in supporting long-term research into fundamental aspects of computing. Such fundamentals provide broad practical benefits, but generally take years to realize. Additionally, the unanticipated results of research are often as important as the anticipated results. Another aspect of government-funded IT research is that it often leads to open standards, something that many perceive as beneficial, encouraging deployment and further investment. Industry, on the other hand, is more inclined to invest in

---

* This is an edited, reformatted and augmented version of Congressional Research Service publication, Report RL33586, dated July 20, 2010.

proprietary products and will diverge from a common standard when there is a potential competitive or financial advantage to do so. Finally, proponents of government support believe that the outcomes achieved through the various funding programs create a synergistic environment in which both fundamental and application-driven research are conducted, benefitting government, industry, academia, and the public. Supporters also believe that such outcomes justify government's role in funding IT R&D, as well as the growing budget for the NITRD Program. Critics assert that the government, through its funding mechanisms, may be picking "winners and losers" in technological development, a role more properly residing with the private sector. For example, the size of the NITRD Program may encourage industry to follow the government's lead on research directions rather than selecting those directions itself.

The President's FY2011 budget request calls for $4.261 billion for the NITRD Program, an increase over the FY2010 request of $3.926 billion (the estimated actual expenditure is $4.305 billion).

On February 8, 2010, the House of Representatives passed H.R. 4061, the Cybersecurity Enhancement Act of 2009, to improve the security of cyberspace by ensuring federal investments in cybersecurity are better focused, more effective, and that research into innovative, transformative technologies is supported. The bill addresses recommendations from the Administration's Cyberspace Policy Review and includes input from four hearings held on cybersecurity during the first session. H.R. 4061 would reauthorize and expand the Cyber Security Research and Development Act (P.L. 107-305). In addition to promoting cybersecurity R&D by the member agencies of the NITRD, the legislation addresses cybersecurity workforce concerns and advances the development of technical standards. H.R. 4061 is a combination of two Committee discussion drafts: the Cybersecurity Research and Development Amendments Act of 2009 and the Cybersecurity Coordination and Awareness Act of 2009. The bill has been referred to the Senate Committee on Commerce, Science, and Transportation.

Overview of the Federal NITRD Program

The federal government has long played a key role in the country's information technology (IT) research and development (R&D) activities. The government's support of IT R&D began because it had an important interest in creating computers that would be capable of addressing the problems and issues the government needed to solve and study. One of the first such problems was planning the trajectories of artillery and bombs; more recently, such problems include simulations of nuclear testing, cryptanalysis, and weather modeling. That interest continues today. Such complexity requires there be adequate coordination to ensure the government's evolving needs (e.g., homeland security) will continue to be met in the most effective manner possible.

NITRD Structure

The Networking and Information Technology Research and Development (NITRD) Program is a collaborative effort in which 13 agencies coordinate and cooperate to help increase the overall effectiveness and productivity of federal IT R&D.[1] Of those 13 members, the majority of funding, in descending order, goes to the National Science Foundation, National Institutes of Health, Department of Energy (DOE) Office of Science, Defense Advanced Research Projects Agency (DARPA), and DOE National Nuclear Security Administration. Dr. Christopher Greer was named as the director of the NITRD Program in October 2007. **Figure 1** illustrates the organizational structure of the NITRD Program.

The National Coordinating Office (NCO) coordinates the activities of the NITRD Program. On July 1, 2005, the NCO became the "National Coordination Office for Networking and Information Technology Research and Development." The Director of the NCO reports to the Director of the White House Office on Science and Technology Policy (OSTP). The NCO supports the Subcommittee on NITRD (also called the NITRD Subcommittee)[2] and the President's Information Technology Advisory Committee (PITAC).[3]

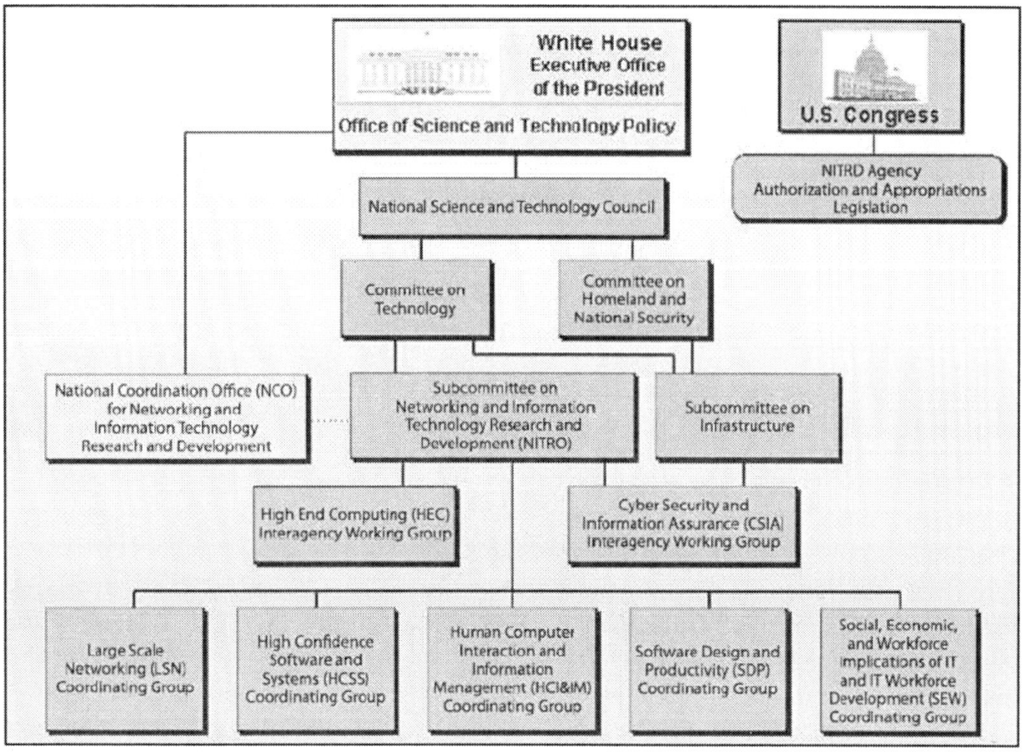

Source: NITRD Program website, http://www.nitrd.gov.

Figure 1. Management Structure of the NITRD Program

- The NITRD Subcommittee provides policy, program, and budget planning for the NITRD Program and is composed of representatives from each of the participating agencies, OSTP, Office of Management and Budget, and the NCO. Two Interagency

Working Groups and five Coordination Groups reporting to the NITRD Subcommittee focus their work in eight Program Component Areas (PCAs).[4]

- The PITAC is composed of representatives of private industry and academia who are appointed by the President. The group provides expert independent advice to the President on the federal role in maintaining U.S. preeminence in advanced IT and works with the NITRD Program agencies and the NITRD Subcommittee.
- The NITRD Program is funded out of each member agency's individual budget, rather than in a single appropriations bill (e.g., NITRD Program activities conducted by the National Institutes of Health (NIH) are funded through the NIH appropriations bill). The program's NCO is not explicitly funded; rather, the NITRD member agencies contribute toward NCO operations.

The NITRD Program has undergone a series of structural changes since its inception in 1991 and both it and the NCO have had a number of different names over the years. When the Program was created in December 1991, it was named the High Performance Computing and Communications (HPCC) Program, and when the NCO was created in September 1992, it was named the NCO for HPCC. The name was changed to the National Coordination Office for Computing, Information, and Communications per the FY1997 Supplement to the President's Budget (also known at that time as the "Blue Book"). The name was changed to the National Coordination Office for Information Technology Research and Development per the FY2001 Blue Book.[5] Most recently, on July 1, 2005, the name was changed to the National Coordination Office for Networking and Information Technology Research and Development. These changes were made to reflect the evolution of the program as it came to encompass a broader range of related topics.

NITRD Funding

The President's FY2011 budget request calls for $4.261 billion for the NITRD Program, an increase over the FY2010 request of $3.926 billion (the estimated actual expenditure is $4.305 billion).

American Recovery and Reinvestment Act of 2009

Under the American Recovery and Reinvestment Act (ARRA) of 2009, five federal agencies report preliminary allocations of $706 million to investments in NITRD research areas (these figures may change). The NITRD agencies will use their ARRA funds to modernize, expand, and upgrade networking and high-end computing infrastructures and facilities for advanced scientific research; expand R&D in cyber security, human-computer interaction and information management, high-confidence software and systems, and software design; and increase investments in education and training for a diverse, highly skilled IT workforce.[6]

NCO, PITAC, and Related Reports and Activities

As explained earlier, the NCO provides technical and administrative support to the NITRD Program, the NITRD Subcommittee, and the PITAC. This includes supporting meetings and workshops and preparing reports. The NCO interacts with OSTP and OMB on NITRD Program and PITAC matters.

National Cyber Leap Year Summit

Between August 17 and 19 of 2009, the NITRD Program, with guidance from OSTP and the Office of the Assistant Secretary for Defense Networks and Information Integration, held a National Cyber Leap Year Summit in Arlington, VA. The Summit gathered commercial and academic innovators for an unconventional exploration of five game-changing strategies in cyber security:

- Basing trust decisions on verified assertions (Digital Provenance)
- Attacks only work once if at all (Moving-target Defense)
- Knowing when we have been had (Hardware-enabled Trust)
- Move from forensics to real-time diagnosis (Nature-inspired Cyber Health)
- Crime does not pay (Cyber Economics)

Participants discussed how to initiate and sustain fundamental cyber security changes within those five strategies. The Summit's outcomes are provided as input to the Administration's cyber security R&D agenda and as strategies for public-private actions to secure the Nation's digital future.[7]

High-Confidence Medical Devices: Cyber-Physical Systems for 21st Century Health Care

This report, published in February 2009, presents the perspectives of the senior scientists of the NITRD Program's High Confidence Software and Systems (HCSS) Coordinating Group (CG), with input from experts from other federal agencies, on the R&D challenges, needs, and strategies for developing and deploying the next generations of high-confidence medical devices, software, and systems.[8] HCSS agencies whose missions are not medical device-specific have found it beneficial to partner in this area because medical device research challenges are similar, if not identical, to those within their purview. Digital technologies are increasingly being assigned high-level control over the monitoring, sensing, actuation, and communications of medical devices—often with human life in the balance. Through this report and associated HCSS-sponsored national workshops, the HCSS agencies are seeking to illuminate fundamental scientific and technical challenges that must be addressed before we can design and build high-confidence devices, software, and systems that operate flawlessly from end to end. The report authors sought to paint the landscape of the evolution of medical device technology and the federal investments that have benefitted medical device R&D over time.

The authors noted a number of key findings:

- Today's medical device architectures are typically proprietary, not interoperable, and rely on professionals to provide inputs and assess outputs; "families" of such devices also tend to be stove-piped and not interoperable with other "families" of devices.
- In the frequent circumstance that a patient is connected to multiple devices at once, such as in an operating room, clinicians now must monitor all devices independently, synthesize data, and act on their observations, which can be affected by stress, fatigue, or other human factors.
- Medical device architecture is beginning to include wired and wireless interfaces to facilitate networked communication of patient data. But ad hoc efforts to aggregate data across devices designed to operate separately can lead to unintended or accidental results.
- The growing interest in such capabilities as home health care services, delivery of expert medical practice remotely (telemedicine), and online clinical lab analysis underscores the central role of advanced networking and distributed communication of medical information in the health systems of the future. Increased R&D focus on the specialized engineering of networked medical device systems is needed.
- Neither past nor current development methods are adequate for the high-confidence design and manufacture of highly complex, interoperable medical device software and systems ("intelligent" prosthetics, minimally invasive surgical devices, implants, "operating room of the future"), which in years to come will likely include nano/bio devices, bionics, or even pure (programmable) biological systems.
- Today's verification and validation (V&V) efforts are driven by system-life-cycle development activities that rely primarily on methods of post-hoc inspection and testing; these approaches are inadequate in the face of the diversity and complexity of components and interactions in emerging medical devices and systems.
- Today, scientific principles and engineering foundations are lacking that could enable both the design and assurance of high-confidence medical device cyber-physical systems.

Based on their findings, the authors drew the following conclusions:

- Clearly, there is a need for rationally designed high-confidence medical device cyber-physical architectures; a strategic focus on R&D in compositional modeling and design is needed to address the open systems needs, respond to technological innovation, and bridge the jointly cyber and physical aspects of this complex systems problem.
- An open research community of academics and medical device manufacturers is needed to create strategies for development of end-to-end, principled, engineering-based design and development tools. Certifying component devices is necessary, but not sufficient; a key area of research needed is the incremental certified composition of certified components.
- Manufacturers will need access to open, formally composable V&V technology that relies on computational models unifying cyber and physical systems to help establish sufficient evidence. A key V&V research challenge is to understand what is meant by

the term "sufficient evidence,"7 its properties, and how this can be accepted in the global economy.

- The HCSS group recommends that a strategic R&D focus on high-confidence networking and IT for the design, implementation, and certification of open medical technologies be undertaken, both to meet the goals of cost-effective, improved patient care and to spur innovation that promotes U.S. leadership in biomedical technology.
- To enable the necessary holistic cyber-physical systems understanding, barriers must fall among the relevant disciplines: medicine, discrete and continuous mathematics of dynamics and control; real-time computation and communication; medical robotics; learning; computational models and the supporting systems engineering design, analysis, and implementation technologies; and formal and algorithmic methods for stating, checking, and reasoning about system properties.
- Incentives are needed to enable effective cooperation between government, industry, and academia to build the underpinning standards and networking and information technology frameworks (e.g., testbeds) for developing open, interoperable medical cyber-physical systems.

Harnessing the Power of Digital Data for Science and Society

This report, published in January 2009, provides a strategy to promote preservation and access to digital scientific data.[9] The report lays out a strategic vision for "a digital scientific data universe in which data creation, collection, documentation, analysis, preservation, and dissemination can be appropriately, reliably, and readily managed, thereby enhancing the return on our nation's research and development investment by ensuring that digital data realize their full potential as catalysts for progress in our global information society." The report includes three key recommendations to pursue this vision. The first is to create an Interagency Subcommittee under NSTC that will focus on goals that are best addressed through continuing broad cooperation and coordination across agencies. The second key element of the strategic framework is for departments and agencies to lay the foundations for agency digital scientific data policy and make the policy publicly available. In laying these foundations, agencies should consider all components of a comprehensive policy to address the full data management life cycle. The third key element is for all agencies to promote a data management planning process for projects that generate scientific data for preservation.

Federal Plan for Advanced Networking Research and Development

This plan, released in September 2008, was developed by the Interagency Task Force on Advanced Networking, established under the NITRD National Science and Technology Council by the Director of the OSTP to provide a strategic vision for future networked environments.[10] The overall conclusions of the Task Force can be summarized as follows:

- Improved networking security and reliability are strategic national priorities;
- New paths to advanced networking are required;
- Federal R&D efforts will support a spectrum of advanced networking capabilities;

- Close cooperation is needed to integrate federal R&D efforts with the full technology development cycle—this cycle includes basic and applied research, and partnerships with researchers, application developers, users, and other stakeholders; and
- Testbeds and prototype networks enable research on network challenges in realistic environments.

The report notes that

> The Internet's phenomenal growth and elasticity have exceeded all expectations. At the same time, we have become captive to the limitations and vulnerabilities of the current generation of networking technologies. Because vital U.S. interests—for example, national defense communications, financial markets, and the operation of critical infrastructures such as power grids—now depend on secure, reliable, highspeed network connectivity, these limitations and vulnerabilities can threaten our national security and economic competitiveness. Research and development to create the next generation of networking technologies is needed to address these threats.
>
> The plan is centered on a vision for advanced networking based on a design and architecture for security and reliability that provides for heterogeneous, anytime-anywhere networking with capabilities such as federation of networks across domains and widely differing technologies; dynamic mobile networking with autonomous management; effective quality of service (QoS) management; support for sensornets; near-realtime autonomous discovery, configuration, and management of resources; and end-to-end security tailored to the application and user.

The report outlines four goal for realizing this vision:

- Provide secure network services anytime, anywhere;
- Make secure global federated networks possible;
- Manage network complexity and heterogeneity; and
- Foster innovation among the federal, research, commercial, and other sectors through development of advanced network systems and technologies.

Leadership Under Challenge: Information Technology R&D in a Competitive World

This August 2007 report assesses global U.S. competitiveness in networking and information technology and provides recommendations aimed at ensuring that the NITRD Program is appropriately focused and implemented. The report makes specific recommendations for federal R&D that would enhance U.S. competitiveness in this economically critical area. In developing the report, the PCAST consulted extensively with experts from industry and academia. The PCAST concluded that while the United States is still in a leadership position, other nations are challenging that lead in a number of areas and that the NITRD Program must focus on visionary research and work with universities to keep the United States at the cutting edge. Some of the report recommendation areas follows:

- Both the U.S. federal government and the private sector need to address the demand for skilled IT professionals, including such steps as updating curricula, increasing fellowships, and simplifying visa processes.

- With respect to the federally funded research portfolio, the NITRD Program should emphasize larger-scale and longer-term, multidisciplinary IT R&D and innovative, higher-risk projects.
- The United States should give priority to R&D in economically important areas, including IT systems connected with and embedded in the physical world, software, use and management of digital data, and advanced Internet capabilities. The PCAST noted that with an annual federal investment of over $3 billion in the NITRD Program, changes in the Program's interagency process to strengthen assessment and planning are needed.[11]

Federal Plan for Cyber Security and Information Assurance Research and Development

In April 2006, the NITRD Subcommittee released its "Federal Plan for Cyber Security and Information Assurance Research and Development."[12] This report sets out a framework for multi-agency coordination of federal R&D investments in technologies that can better secure the interconnected computing systems, networks, and information that together make up the U.S. IT infrastructure. The plan outlines strategic objectives for coordinated federal R&D in cyber security and information assurance (CSIA) and presents a broad range of CSIA R&D technical topics, identifying those topics that are multi-agency technical and funding priorities. The plan's findings and recommendations address R&D priority setting, coordination, fundamental R&D, emerging technologies, roadmapping, and metrics.

NSA Superconducting Technology Assessment

In August 2005, NSA released its "Superconducting Technology Assessment"[13] as part of its participation in the High-End Computing PCA of the NITRD Program. NSA had been concerned about projected limitations of conventional technology and wanted to explore possible alternatives to meet its future mission-critical computational needs. This report presented the results of the technology assessment, which found the following:

- Government investment is necessary, because private industry currently has no compelling financial reason to develop alternative technologies for mainstream commercial applications.
- With aggressive federal investment (estimated between $372 and $437 million over five years), by 2010 next generation technologies would be sufficiently mature to allow the initiation of the design and construction of an operational petaflops[14]-scale system.
- Although significant risks exist, the panel has developed a roadmap that identifies the needed technology developments with milestones and demonstration vehicles.

Computational Science: Ensuring America's Competitiveness

In June 2005, the PITAC released "Computational Science: Ensuring America's Competitiveness."[15] The report identified obstacles to progress in this field, including "rigid disciplinary silos in academia that are mirrored in federal research and development agency organizational structures." According to the report, these "silos stifle the development of multidisciplinary research and educational approaches essential to computational science." The report recommends the following:

- Both academia and government fundamentally change their organizational structures so that they promote and reward collaborative research.
- The National Science and Technology Council commission the National Academies to convene one or more task forces to develop and maintain a multi-decade roadmap for computational science, with a goal of assuring continuing U.S. leadership in science, engineering, and the humanities.
- The federal government establish national software sustainability centers to harden, document, support, and maintain long-term vital computational science software.
- The federal government provide long-term support for computational science community data repositories. These should include defined frameworks, metadata structures, algorithms, data sets, applications, and review and validation infrastructure. It should also require funded researchers to deposit their data and research software in these repositories or with other approved access providers.
- The federal government provide long-term funding for national high-end computing centers at levels sufficient to ensure the regularly scheduled deployment and operation of the fastest and most capable high-end computing systems that address the most demanding computational problems.
- The federal government implement coordinated, long-term computational science programs that include funding for interconnecting the software sustainability centers, national data and software repositories, and national high-end leadership centers with the researchers who use those resources.
- The federal government should rebalance its R&D investments to: (a) create a new generation of well-engineered, scalable, easy-to-use software suitable for computational science that can reduce the complexity and time to solution for today's challenging scientific applications and can create accurate simulations that answer new questions; (b) design, prototype, and evaluate new hardware architectures that can deliver larger fractions of peak hardware performance on scientific applications; and (c) focus on sensor-and data-intensive computational science applications in light of the explosive growth of data.

Cyber Security: A Crisis of Prioritization

In February 2005, the PITAC released "Cyber Security: A Crisis of Prioritization."[16] That report outlined four key findings and recommendations on how the federal government could "foster new architectures and technologies to secure the Nation's IT infrastructure." Specifically, the PITAC urged the government to

- significantly increase support for fundamental research in civilian cyber security in 10 priority areas;
- intensify federal efforts to promote the recruitment and retention of cyber security researchers and students at research universities;
- increase support for the rapid transfer of federally developed cybersecurity technologies to the private sector; and

- strengthen the coordination of federal cybersecurity R&D activities.

Also in February 2005, the NCO released the FY2006 Supplement to the President's Budget.[17] The supplement provides a brief technical outline of the FY2006 budget request for the NITRD Program. The FY2007 Supplement has not yet been released.

NITRD Enabling and Governing Legislation

The NITRD Program is governed by two laws. The first, the High-Performance Computing Act of 1991, P.L. 102-194,[18] expanded federal support for high-performance computing R&D and called for increased interagency planning and coordination. The second, the Next Generation Internet Research Act of 1998, P.L. 105-305,[19] amended the original law to expand the mission of the NITRD Program to cover Internet-related research, among other goals.

High-Performance Computing Act of 1991

This law was the original enabling legislation for what is now the NITRD Program. Among other requirements, it called for the following:

- Setting goals and priorities for federal high-performance computing research, development, and networking.
- Providing for the technical support and research and development of high-performance computing software and hardware needed to address fundamental problems in science and engineering.
- Educating undergraduate and graduate students.
- Fostering and maintaining competition and private sector investment in high-speed data networking within the telecommunications industry.
- Promoting the development of commercial data communications and telecommunications standards.
- Providing security, including protecting intellectual property rights.
- Developing accounting mechanisms allowing users to be charged for the use of copyrighted materials.

This law also requires an annual report to Congress on grants and cooperative R&D agreements and procurements involving foreign entities.[20]

Next Generation Internet Research Act of 1998

This law amended the High-Performance Computing Act of 1991. The act had two overarching purposes. The first was to authorize research programs related to high-end computing and computation, human-centered systems, high confidence systems, and education, training, and human resources. The second was to provide for the development and coordination of a comprehensive and integrated U.S. research program to focus on (1) computer network infrastructure that would promote interoperability among advanced federal computer networks, (2) economic high-speed data access that does not impose a "geographic penalty." and (3) flexible and extensible networking technology.

Context of Federal Technology Funding

In the early 1990s, Congress recognized that several federal agencies had ongoing high-performance computing programs,[21] but no central coordinating body existed to ensure long-term coordination and planning. To provide such a framework, Congress passed the High-Performance Computing Program Act of 1991 to improve the interagency coordination, cooperation, and planning of agencies with high performance computing programs.

In conjunction with the passage of the act, OSTP released, "Grand Challenges: High-Performance Computing and Communications." That document outlined an R&D strategy for high-performance computing and communications and a framework for a multi-agency program, the HPCC Program.

The NITRD Program is part of the larger federal effort to promote fundamental and applied IT R&D. The government sponsors such research through a number of channels, including

- federally funded research and development laboratories, such as Lawrence Livermore National Laboratory;
- single-agency programs;
- multi-agency programs, including the NITRD Program, but also programs focusing on nanotechnology R&D and combating terrorism;
- funding grants to academic institutions; and
- funding grants to industry.

In general, supporters contend that federal funding of IT R&D has produced positive results. In 2003, the Computer Science and Telecommunications Board (CSTB) of the National Research Council (NRC) released a "synthesis report" based on eight previously released reports that examined "how innovation occurs in IT, what the most promising research directions are, and what impacts such innovation might have on society."[22] One of the most significant of the CSTB's observations was that the unanticipated results of research are often as important as the anticipated results. For example, electronic mail and instant messaging were by-products of [government-funded] research in the 1960s that was aimed at making it possible to share expensive computing resources among multiple simultaneous interactive users.

Additionally, the report noted that federally funded programs have played a crucial role in supporting long-term research into fundamental aspects of computing. Such "fundamentals" provide broad practical benefits, but generally take years to realize. Furthermore, supporters state that the nature and underlying importance of fundamental research makes it less likely that industry would invest in and conduct more fundamental research on its own. As noted by the CSTB, "companies have little incentive to invest significantly in activities whose benefits will spread quickly to their rivals."[23] Further, in the Board's opinion:

> government sponsorship of research, especially in universities, helps develop the IT talent used by industry, universities, and other parts of the economy. When companies create products using the ideas and workforce that result from federally-sponsored research, they repay the nation in jobs, tax revenues, productivity increases, and world leadership.[24]

Another aspect of government-funded IT R&D is that it often leads to open standards, something that many perceive as beneficial, encouraging deployment and further investment. Industry, on the other hand, is more likely to invest in proprietary products and will diverge from a common standard if it sees a potential competitive or financial advantage; this has happened, for example with standards for instant messaging.[25]

Finally, proponents of government R&D support believe that the outcomes achieved through the various funding programs create a synergistic environment in which both fundamental and application-driven research are conducted, benefitting government, industry, academia, and the public. Supporters also believe that such outcomes justify government's role in funding IT R&D, as well as the growing budget for the NITRD Program.

Critics assert that the government, through its funding mechanisms, may be setting itself up to pick "winners and losers" in technological development, a role more properly residing with the private sector.[26] For example, the size of the NITRD Program may encourage industry to follow the government's lead on research directions rather than selecting those directions itself.

Overall, CSTB states that, government funding appears to have allowed research on a larger scale and with greater diversity, vision, and flexibility than would have been possible without government involvement.[27]

Issues for Congress

Federal IT R&D is a multi-dimensional issue, involving many government agencies working together towards shared and complementary goals. Most observers believe that success in this arena requires ongoing coordination among government, academia, and industry.

Through hearings, the House Committee on Science has been investigating issues related to U.S. competitiveness in high-performance computing and the direction the IT R&D community has been taking. Those issues and others remain salient and may merit further investigation if the United States is to maintain a comprehensive IT R&D policy. Included among the possible issues Congress may wish to pursue are: the United States' status as the global leader in high-performance computing research; the apparent bifurcation of the federal IT R&D research agenda between grid computing and supercomputing capabilities; the possible over-reliance on commercially available hardware to satisfy U.S. research needs; and the potential impact of deficit cutting on IT R&D funding.

Many Members of Congress as well as those in the research community have expressed concern over whether the United States is maintaining its position as the global leader in high-performance computing R&D. That concern was highlighted in 2003 when Japan briefly surpassed the United States in possessing the fastest and most efficient supercomputer in the world.[28] While this was a reason for some concern, it was also viewed by some as an indicator of how the United States' research agenda had become bifurcated, with some in the R&D community focusing on traditional supercomputing capabilities, and others focusing more on cluster computing or grid computing. Each type of computing has its advantages, based on its application. Stand-alone supercomputers are often faster and are generally used to work on a specific problem. For example, cryptanalysis and climate modeling applications require significant computing power and are best accomplished using specialized, stand-alone computers. Cluster computing, however, allows the use of commercially available hardware, which helps contain costs. The cluster configuration is useful for applications in which a problem can be broken into smaller independent components.[29] Therefore, one possible course for Congress could be to monitor closely the work that was begun by the High-End Computing Revitalization Task Force and is now being performed by the NITRD Program's High-End Computing Interagency Working Group and provide ongoing feedback and guidance.

Without a clear plan as to how to proceed, pursuing two disparate research agendas (with goals that could be viewed as being at odds with each other) could split the research community further, damaging its ability to provide leadership in either area. The NITRD Program already is working on a "roadmap" for future directions in supercomputing; therefore, one possible course for Congress at this time would be to monitor closely the work of the High-End Computing Revitalization Task Force and provide input or a more visible forum for discussion (i.e., additional hearings involving task force participants). Congress may wish to conduct its own inquiry into the debate over grid versus stand-alone computing.

Another issue is whether the United States is relying too heavily on commercially available hardware to satisfy its R&D needs. While use of computers designed for mass-market commercial applications can certainly be a part of a successful high-end computing R&D plan, Congress may wish to monitor how this reliance may be driving the new emphasis on grid computing.

As noted earlier, critics of IT R&D funding often state that industry should conduct more fundamental R&D on their own, without government backing, and that fiscal restraint dictates that less funding should be made available. Conversely, supporters of government funding would point out that IT R&D has a very long cycle from inception to application and that any reductions in funding now could have a significant negative impact for many years to come in terms of innovation and training of researchers. Therefore, Congress may monitor and assess the potential impact of deficit-cutting plans on progress in IT R&D.

ACTIVITY IN THE 111TH CONGRESS

The most recent legislative activity regarding the NITRD Program was the February 8, 2010, passage H.R. 4061, the Cybersecurity Enhancement Act of 2009, by the House of Representatives (H.Rept. 111-405). The bill is intended to improve the security of cyberspace by ensuring federal investments in cybersecurity are better focused, more effective, and that research into innovative, transformative technologies is supported. It also addresses recommendations from the Administration's Cyberspace Policy Review and includes input from four hearings held on cybersecurity during the first session. H.R. 4061 would reauthorize and expand the Cyber Security Research and Development Act (P.L. 107-305). In addition to promoting cybersecurity R&D by the member agencies of the NITRD, the legislation addresses cybersecurity workforce concerns and advances the development of technical standards. H.R. 4061 is a combination of two Committee discussion drafts: the Cybersecurity Research and Development Amendments Act of 2009 and the Cybersecurity Coordination and Awareness Act of 2009. The bill has been referred to the Senate Committee on Commerce, Science, and Transportation.

Four other bills have also been introduced that relate to the NITRD Program.

H.R. 1, the American Recovery and Reinvestment Act of 2009 (P.L. 111-5), was signed into law on February 17, 2009. Prior to being signed by the President, H.R. 1 was amended to include two other related bills, H.R. 598 and H.R. 629.

The fourth bill, H.R. 2020, the Networking and Information Technology Research and Development Act of 2009, was introduced by Representative Bart Gordon on April 22, 2009; it was passed by the House of Representatives and referred to the Senate Committee on Commerce, Science, and Transportation on May 13, 2009 (see H.Rept. 111-102). The purpose of this bill is to strengthen the planning and coordination mechanisms of the NITRD Program and to update the research content of the program. The legislation implements a number of recommendations made in a recent PCAST assessment of the program.

End Notes

[1] The members of the NITRD Program, as listed in the FY2006 Supplement to the President's Budget, are: Agency for Healthcare Research and Quality (AHRQ); Defense Advanced Research Projects Agency (DARPA); Office of the Secretary of Defense, Defense Research & Engineering, and the DOD service research organizations; Department of Energy, National Nuclear Security Administration (DOE/NNSA); Department of Energy, Office of Science (DOE/SC); Department of Homeland Security (DHS); Environmental Protection Agency (EPA); National Aeronautics and Space Administration (NASA); National Institutes of Health (NIH); National Institute of Standards and Technology (NIST); National Oceanic and Atmospheric Administration (NOAA); National Security Agency (NSA); and National Science Foundation (NSF). The history of agency participation can be found at http://www.nitrd.gov/about/history/agency-participants.pdf.

[2] The NITRD Subcommittee was previously called the Interagency Working Group for IT R&D (IWG/IT R&D).

[3] The PITAC was established on February 11, 1997, to provide the President, OSTP, and the federal agencies involved in IT R&D with guidance and advice on all areas of high performance computing, communications, and information technologies. Representing the research, education, and library communities and including network providers and representatives from critical industries, the committee advises the Administration's effort to accelerate development and adoption of information technologies. Additional information about the PITAC is available at http://www.nitrd.gov/pitac. The most recent PITAC Executive Order expired on June 1, 2005.

[4] The eight PCAs are (1) *High-End Computing Infrastructure and Applications (HEC I&A)*—to extend the state of the art in high-end computing systems, applications, and infrastructure; (2) *High-End Computing R&D (HEC R&D)*—to optimize the performance of today's high-end computing systems and develop future generations of high-end computing systems; (3) *Cyber Security and Information Assurance*—to perform fundamental and applied R&D to improve the security and assurance of information systems; (4) *Human Computer Interaction and Information Management (HCI&IM)*—to develop new user interaction technologies, cognitive systems, information systems, and robotics that benefit humans; (5) *Large Scale Networking (LSN)*—to develop leading-edge network technologies, services, and techniques to enhance performance, security, and scalability; (6) *Software Design and Productivity (SDP)*—to advance concepts, methods, techniques, and tools that improve software design, development, and maintenance to produce more usable, dependable and cost-effective software-based systems; (7) *High Confidence Software and Systems (HCSS)*—to develop the scientific foundations and IT to achieve affordable and predictable high levels of safety, security, reliability, and survivability, especially in U.S. national security and safety-critical systems; and (8) n *Social, Economic, and Workforce Implications of IT and IT Workforce Development (SEW)*—to study the impact of IT on people

and social and economic systems; develop the IT workforce; and develop innovative IT applications in education and training. Additional information about the program component areas is available at http://www.nitrd.gov/subcommittee/index.html. HEC R&D and HEC I&A are both covered by the HEC Interagency Working Group. A diagram illustrating the evolution of the PCAs, 1992-present, is available at http://www.nitrd.gov/ about/history/new-pca-names.pdf.

[5] That change was effective October 2000.

[6] Ibid.

[7] The reports from the summit are available at http://www.qinetiq-na.com/.

[8] This report is available online at http://www.nitrd.gov/About/MedDevice-FINAL1-web.pdf.

[9] This report is available online at http://www.nitrd.gov/About/Harnessing_Power_Web.pdf.

[10] This report is available at http://www.nitrd.gov/pubs/ITFAN-FINAL.pdf.

[11] This report responds to reporting requirements of the High-Performance Computing Act of 1991 (P.L. 102-194) and the Next Generation Internet Research Act of 1998 (P.L. 105-305). The laws call for a President's Information Technology Advisory Committee (PITAC) to assess periodically what is now known as the NITRD Program. Executive Order 13385, signed on September 29, 2005, assigned the PITAC's responsibilities to PCAST. This report is available at http://www.nitrd.gov/pcast/reports/PCAST-NIT-FINAL.pdf.

[12] This report is available at http://www.nitrd.gov/pubs/csia/csia_federal_plan.pdf.

[13] This report is available at http://www.nitrd.gov/pubs/nsa/sta.pdf.

[14] In computing, "flops" or "FLOPS" is an abbreviation of Floating Point Operations Per Second. This is used as a measure of a computer's performance, especially in fields of scientific calculations that make heavy use of floating point calculations. A petaflops-scale machine operates at 10^{15} flops.

[15] This report is available at http://www.nitrd.gov/pitac/reports/20050609_computational/computational.pdf.

[16] This report is available at http://www.nitrd.gov/pitac/reports/20050301_cybersecurity/cybersecurity.pdf.

[17] This report is available at http://www.nitrd.gov/pubs/2006supplement.

[18] High Performance Computing Act of 1991, P.L. 102-194, 15 U.S.C. 5501, 105 Stat. 1595, December 9, 1991. The full text of this law is available at http://www.nitrd.gov/congressional/laws/pl_102-194.html.

[19] Next Generation Internet Research Act of 1998, P.L. 105-305, 15 U.S.C. 5501, 112 Stat. 2919, October 28, 1998. The full text of this law is available at http://www.nitrd.gov/congressional/laws/pl_h_105-305.html.

[20] The first report mandated information on the "Supercomputer Agreement" between the United States and Japan be included in this report. A separate one-time only report was required on network funding, including user fees, industry support, and federal investment.

[21] "High-performance" computing is a term that encompasses both "supercomputing" and "grid computing." In general, high-performance computers are defined as stand-alone or networked computers that can perform "very complex computations very quickly." Supercomputing involves a single, stand-alone computer located in a single location. Grid computing involves a group of computers, in either the same location or spread over a number of locations, that are networked together (e.g., via the Internet or a local network). House of Representatives, Committee on Science, *Supercomputing: Is the United States on the Right Path* (Hearing Transcript), http://commdocs.house.gov/committees/ science/hsy88231.000/hsy88231_0f.htm, 2003, pp. 5-6.

[22] National Research Council, *Innovation in Information Technology*, 2003, p. 1. This report discusses all federal funding for R&D, not only the NITRD Program.

[23] Ibid, p. 4.

[24] Ibid, p. 4.

[25] Ibid, p. 18.

[26] Cato Institute, *Encouraging Research: Taking Politics Out of R&D*, September 13, 1999, http://www.cato.org/ pubs/ wtpapers/990913catord.html.

[27] National Research Council, *Innovation in Information Technology*, 2003, p. 22.

[28] House of Representatives, Committee on Science, *Supercomputing: Is the United States on the Right Path?* (Hearing Transcript), http://commdocs.house.gov/committees/science/hsy88231.000/hsy88231_0f.htm, 2003, p. 13.

[29] Ibid, p. 6-7.

In: Technology Policies and Issues for the U.S. Government ISBN: 978-1-61209-501-1
Editors: Danielle A. Godden and Mason C. Andrews

Chapter 8

THE NATIONAL NANOTECHNOLOGY INITIATIVE: OVERVIEW, REAUTHORIZATION AND APPROPRIATION ISSUES*

John F. Sargent, Jr.

SUMMARY

Nanotechnology—a term encompassing the science, engineering, and applications of submicron materials—involves the harnessing of unique physical, chemical, and biological properties of nanoscale substances in fundamentally new and useful ways. The economic and societal promise of nanotechnology has led to substantial and sustained investments by governments and companies around the world. In 2000, the United States launched the world's first national nanotechnology program. From FY2001 through FY2010, the federal government invested approximately $12.4 billion in nanoscale science, engineering, and technology through the U.S. National Nanotechnology Initiative (NNI). U.S. companies and state governments have invested billions more. President Obama has requested an additional $1.8 billion in NNI funding for FY2011. As a result of this focus and these investments, the United States has, in the view of many experts, emerged as a global leader in nanotechnology. However, the competition for global leadership in nanotechnology is intensifying as countries and companies around the world increase their investments.

Nanotechnology's complexity and intricacies, early stage of development (with commercial payoff possibly years away for many potential applications), and broad scope of potential applications engender a wide range of public policy issues. Maintaining U.S. technological and commercial leadership in nanotechnology poses a variety of technical and policy challenges, including development of technologies that will enable commercial scale

* This is an edited, reformatted and augmented version of Congressional Research Service publication, Report RL34401, dated May 13, 2010.

manufacturing of nanotechnology materials and products; environmental, health, and safety (EHS) concerns; and maintenance of public confidence in its safety.

Congress established programs, assigned responsibilities, and initiated research and development (R&D) related to these issues in the 21st Century Nanotechnology Research and Development Act of 2003 (P.L. 108-153). While many provisions of this act have no sunset provision, FY2008 was the last year of agency authorizations included in the act. Legislation to amend and reauthorize the act was introduced in the House (H.R. 5940, 110th Congress) and the Senate (S. 3274, 110th Congress) in the 110th Congress. Both bills were titled the National Nanotechnology Initiative Amendments Act of 2008. The House passed H.R. 5940 by a vote of 407-6; the Senate did not act on S. 3274. In January 2009, H.R. 554, the National Nanotechnology Initiative Amendments Act of 2009, was introduced in the 111th Congress. The act contains essentially the same provisions as H.R. 5940 (110th Congress). In February 2009, the House passed the bill by voice vote under a suspension of the rules. The bill was referred to the Senate Committee on Commerce, Science, and Transportation; no further action has been taken. On May 7, 2010, the House Committee on Science and Technology reported the America COMPETES Reauthorization Act of 2010 (H.R. 5116) which includes, as Title I, Subtitle A, the National Nanotechnology Initiative Amendments Act of 2010. Provisions of this subtitle are nearly identical to the provisions of H.R. 554.

Proponents of the NNI assert that nanotechnology is one of the most important emerging and enabling technologies and that U.S. competitiveness, technological leadership, national security, and societal interests require an aggressive approach to the development and commercialization of nanotechnology. Critics of the NNI voice concerns that reflect disparate underlying beliefs. Some critics assert that the government is not doing enough to move technology from the laboratory into the marketplace. Others argue that the magnitude of the public investment may skew what should be market-based decisions in research, development, and commercialization. Still other critics say that the inherent risks of nanotechnology are not being addressed in a timely or effective manner.

Introduction

Nanotechnology has been an issue of interest to Congress for a number of years, coming into focus in 2000 with the launch of the U.S. National Nanotechnology Initiative (NNI) by President Clinton in his FY2001 budget request to Congress. From FY2001 through FY2010, Congress appropriated approximately $12.4 billion for nanotechnology research and development (R&D).[1] These efforts have been directed at advancing understanding and control of matter at the nanoscale,[2] where the physical, chemical, and biological properties of materials differ in fundamental and useful ways from the properties of individual atoms or bulk matter.[3]

Nanotechnology: A Description

The term "nanotechnology" is often used as an all-encompassing term for nanoscale science, engineering, and technology. Nanotechnology is the understanding and control of matter at dimensions of roughly 1 to 100 nanometers, the size-scale between individual atoms and bulk materials, where unique phenomena enable novel applications. A nanometer is one-billionth of a meter, or about the width of 10 hydrogen atoms arranged side-by-side in a line. Nanotechnology involves imaging, measuring, modeling, and manipulating matter at this size-scale.

At the nanoscale, the physical, chemical, and biological properties of materials can differ in fundamental and useful ways from the properties of individual atoms and molecules or bulk matter. Nanotechnology R&D is directed toward understanding and creating improved materials, devices, and systems that exploit these new properties.

Physicist Richard Feynman's remarks at the 1959 annual meeting of the American Physical Society are often cited as the first articulation of and vision for nanotechnology. Though he did not use the term nanotechnology in this speech, he spoke of controlling matter at the nanoscale and creating atomic-level machines, positing some of the applications that doing so might enable.

Source: The National Nanotechnology Initiative Strategic Plan, 2004, Nanoscale Science, Engineering, and Technology Subcommittee, National Science and Technology Council, The White House, December 2004.

The development and application of nanotechnology—more fully explained below—across a wide array of products and industries holds the potential for significant economic and societal benefits. To capture these benefits, the United States will have to effectively address a variety of technical and policy challenges that stand as potential barriers to commercialization, including environmental, health, and safety (EHS) concerns and their implications for workplace, environmental, food, and drug regulations; development of standards, reference materials, and consistent nomenclature; development of new measurement methods and tools; effective technology transfer to the private sector; protection of intellectual property; availability, affordability, and patience of investment capital; ethical, legal, and societal concerns; public understanding, support, and acceptance; and development of a world-class scientific and technical nanotechnology workforce.

This report provides an overview of nanotechnology, the National Nanotechnology Initiative, possible reauthorization of the 21st Century Nanotechnology Research and Development Act of 2003 (P.L. 108-153), and appropriations issues.

OVERVIEW

The economic and societal promise of nanotechnology has led to involvement and investments by governments and companies around the world. In 2000, the United States became the first nation to establish a formal, national initiative to advance nanoscale science, engineering, and technology—the National Nanotechnology Initiative. Since then, Congress has appropriated approximately \$12.4 billion in nanoscale science, engineering, and technology through the NNI. U.S. companies and state governments have invested billions more.

As a result of this focus and these investments, the United States has emerged as a global leader in nanotechnology. However, the competition for global leadership is intensifying as foreign investments in nanoscale science, engineering, and technology increase. Other nations have followed the U.S. lead and established their own national nanotechnology programs, each with varying degrees of investment, foci, and support for industrial applications and commercialization. Today, almost every nation that supports R&D has a national-level nanotechnology program.

Global public investments in nanotechnology in 2009 alone were estimated to be $9.8 billion, with cumulative global public investments through 2009 reaching approximately $50 billion.[4] In 2005, an assessment of the NNI by the President's Council of Advisors on Science and Technology (PCAST) reported that the United States leads all other nations in public investments in nanotechnology R&D, accounting for about a quarter of global annual public investments.[5] An April 2009 report by Cientifica, a privately-held nanotechnology business analysis and consulting firm, states that the U.S. share of global public nanotechnology funding has fallen to 19% in 2009 from 31% in 2004, calculated on a currency exchange rate basis, behind Russia (23%) and the European Union (EU) States (collectively, 27%).[6] The situation appears even less favorable when the share of investment is calculated on a purchasing power parity (PPP) basis (which takes into account the price of goods and services in each nation). Using a PPP approach, Cientifica concluded in its report that the U.S. share of global public nanotechnology investments is 16%, less than Russia (25%), and the EU States (27%), and the same as China (16%).[7]

Global investments in nanotechnology already have begun to yield economic benefits as products incorporating nanotechnology enter the marketplace. These products are estimated to have produced $147 billion in revenues in 2007 (including $59 billion in the United States, $47 billion in Europe, $31 billion in Asia/Pacific, and $9 billion in other nations).[8] By tapping the unique properties that emerge at the nanoscale, proponents maintain that nanotechnology holds the potential for products that could transform existing industries and create new ones, clean and protect the environment, extend and improve the quality of our lives, and strengthen national security. Most nanotechnology products currently on the market—such as faster computer processors, higher density memory devices, lighter-weight auto parts, stain-resistant clothing, antibiotic bandages, cosmetics, and clear sunscreen—are evolutionary in nature, offering incremental improvements in characteristics such as performance, aesthetics, cost, size, and weight.

Evolutionary nanotechnology products, however, represent only a small fraction of what many see as the substantial longer-term economic and societal promise of nanotechnology. One estimate projects nanotechnology product revenues will reach $3.1 trillion by 2015,[9] while another estimate projects revenues will reach $2.95 billion by 2015, of which almost half will come from semiconductors.[10, 11]

Many nanotechnology advocates—including business executives, scientists, engineers, medical professionals, and venture capitalists—assert that in the longer term, nanotechnology, especially in combination with information technology, biotechnology, and the cognitive sciences, may deliver revolutionary advances, including:

- new prevention, detection, and treatment technologies that could reduce substantially death and suffering from cancer and other deadly illnesses;[12]
- new organs to replace damaged or diseased ones;[13]
- contact lenses, skin patches, and glucose-sensing tattoos that monitor diabetics' blood sugar levels and warn when too high or low;[14]
- clothing that protects against toxins and pathogens;[15]

- clean, inexpensive, renewable power through energy creation, storage, and transmission technologies;[16]
- inexpensive, portable water purification systems that provide universal access to safe water;[17]
- energy efficient, low-emission "green" manufacturing systems;[18]
- high-density memory systems capable of storing the entire Library of Congress collection on a device the size of a sugar cube;[19]
- agricultural technologies that increase crop yield and improve nutritional value, reducing global hunger and malnutrition;[20]
- self-repairing materials;[21]
- powerful, small, inexpensive sensors that can warn of minute levels of toxins and pathogens in air, soil, or water;[22] and
- decontaminated industrial sites through environmental remediation.[23]

While some applications of nanotechnology have proven market-ready, much fundamental research remains ahead, including efforts to advance understanding of nanoscale phenomena; characterize nanoscale materials; understand how to control and manipulate nanoscale particles; develop instrumentation and measurement methods; and understand how nanoscale particles interact with humans, animals, plants, and the environment. In addition, several federal agencies—such as the Departments of Defense, Energy, and Homeland Security—see the potential for nanotechnology to help address mission requirements. Historically, the federal government has played a central role in funding these types of research and development activities.

Though federal nanoscale science, engineering, and technology R&D had been underway for over a decade, the NNI was first initiated as a Presidential technology initiative in 2000.[24] The original participating agencies were the National Science Foundation (NSF), the Department of Defense (DOD), the Department of Energy (DOE), the Department of Commerce's (DOC) National Institute of Standards and Technology (NIST), the National Aeronautics and Space Administration (NASA), and the Department of Health and Human Services' National Institutes of Health (NIH). In 2009, 25 agencies participated in the NNI, including 13 that received appropriations to conduct and/or fund nanotechnology R&D. Since its first year of funding in FY2001, the NNI's annual appropriations have grown nearly four-fold to an estimated $1.8 billion in FY2010.[25]

In 2003, Congress provided a statutory foundation for some of the activities of the NNI through the 21st Century Nanotechnology Research and Development Act of 2003 (P.L. 108-153). The act established a National Nanotechnology Program (NNP) and provided authorizations for a subset of the NNI agencies, namely the NSF, DOE, NASA, NIST, and Environmental Protection Agency (EPA).[26] The act, however, did not address the participation of several agencies that fund nanotechnology R&D under the NNI, including DOD, NIH, and the Department of Homeland Security (DHS). Nevertheless, coordination of nanotechnology R&D activities across all NNI funding agencies continues under the National Science and Technology Council's (NSTC's) Nanoscale Science, Engineering, and Technology (NSET) subcommittee.[27] According to the NSET subcommittee's 2004 NNI Strategic Plan, "For continuity and to capture this broader participation, the coordinated federal activities as a whole will continue to be referred to as the National Nanotechnology

Initiative." Accordingly, the functions and activities established under the act are incorporated into the Executive Branch's implementation of the NNI.

The thrust of the NNI has primarily been the development of fundamental scientific knowledge through basic research. Investments at mission agencies, such as DOD, have supported nanotechnology applications development for which they are a primary customer. Other investments have supported infrastructural technologies. For example, NIST has contributed to developing tools and standards that enable measurement and control of matter at the nanoscale, thereby supporting the conduct of R&D and the ability to manufacture nanoscale materials and products. As understanding of nanotechnology has matured, the NNI has worked with a variety of industry organizations to facilitate the movement of research results from the laboratory bench to the marketplace in fields as disparate as semiconductors, chemicals, energy, concrete, and forest products.

The NNI agencies also have begun to address research needs and regulatory issues related to environmental, health, and safety issues, as well as issues such as public understanding and workforce education and training. The NNI agencies actively engage in a variety of international fora, such as the Organization for Economic Cooperation and Development (OECD) and the International Standards Organization (ISO), to cooperatively address nanotechnology issues related to EHS, metrology[28] and standards, nomenclature, and nanoscale materials characterization.

Maintaining U.S. leadership poses a variety of technical, economic, and policy challenges, including:

- safeguarding the environment and ensuring human health and safety;
- creating the standards, reference materials, nomenclature, methods, and tools for metrology to enable the manufacturing of nanoscale materials and products;
- developing a world-class scientific and technical nanotechnology workforce;
- translating research results into products, including effective technology transfer to the private sector;
- understanding public perceptions and attitudes and fostering public understanding;
- addressing ethical, legal and societal implications;
- protecting intellectual property;
- securing investment capital for early-stage research, development, and commercialization; and
- fostering and facilitating international cooperation and coordination.

Proponents of the NNI assert that nanotechnology is one of the most important emerging and enabling technologies[29] and that U.S. competitiveness, technological leadership, national security, and societal interests require an aggressive approach to the development and commercialization of nanotechnology. Critics of the NNI hold a variety of competing views, asserting that government is not doing enough, is doing too much, or is moving too quickly.

Some in industry have criticized the NNI for being overly focused on basic research and not being aggressive enough in moving NNI-funded R&D out of government and university laboratories and into industry. Others in industry have criticized the federal government for not providing mechanisms to help advance nanotechnology R&D to the point where it becomes economically viable for venture capitalists, corporations, and other investors to create products and bring them to market. Some refer to this gap as the "valley of death."[30] Still others in industry have criticized the NNI for not adequately supporting the development of metrology, standards, equipment, and processes necessary to manufacture nanotechnology materials, products, and systems at a commercial scale.

Conversely, supporters of industry-driven market investments contend that extensive government support for nanotechnology may supplant the judgment of the marketplace by picking "winners and losers" in technological development. For example, the size and directions of the NNI investments may encourage industry to follow the government's lead rather than independently selecting R&D directions itself or, alternatively, may result in the promotion of a less effective technology path over a more effective one. These supporters also assert that federal government funding of scientific research is often wasteful, driven by political considerations and not scientific merit.[31]

Some non-governmental organizations (NGO) are critical of nanotechnology for its potential adverse impacts on human health and safety and on the environment. They assert that the government is pushing ahead too quickly in developing nanotechnology and encouraging its commercialization and use without adequately investing in research focused on understanding and mitigating negative EHS implications.[32] They argue that the very characteristics that make nanotechnology promising also present significant potential risks to human health and safety and the environment. Some of these critics argue for application of the "precautionary principle," which holds that regulatory action may be required to control potentially hazardous substances even before a causal link has been established by scientific evidence.[33] At least one NGO has called for a moratorium on nanotechnology R&D and new commercial products incorporating synthetic nanoparticles.[34]

National Nanotechnology Initiative

The National Nanotechnology Initiative is an interagency program that coordinates federal nanoscale science, engineering, and technology R&D activities and related efforts among participating agencies.

Vision and Goals

The National Science and Technology Council (NSTC) has stated the following vision for the NNI:

> A future in which the ability to understand and control matter on the nanoscale leads to a revolution in technology and industry. The NNI will expedite the discovery, development, and deployment of nanotechnology in order to achieve responsible and sustainable economic benefits, to enhance the quality of life, and to promote national security.[35]

To achieve its vision, the NNI has established four goals: maintain a world-class R&D program aimed at realizing the full potential of nanotechnology; facilitate transfer of new technologies into products that provide economic growth, jobs, and other public benefits;

develop educational resources, a skilled workforce, and the supporting infrastructure and tools to advance nanotechnology; and support responsible development of nanotechnology.[36]

History

Attempts to coordinate federal nanoscale R&D began in November 1996, as staff members from several agencies met regularly to discuss their plans and programs in nanoscale science and technology. This group continued informally until September 1998, when it was designated as the Interagency Working Group on Nanotechnology (IWGN) under the NSTC. In August 1999, IWGN completed its first draft of a plan for an initiative in nanoscale science and technology, which was subsequently approved by the President's Council of Advisors on Science and Technology (PCAST) and the White House Office of Science and Technology Policy (OSTP).[37]

In his 2001 budget submission to Congress, then-President Clinton raised nanotechnology-related research to the level of a federal initiative, officially referring to it as the National Nanotechnology Initiative.[38]

Legislative Approach

Congress has played a central role in the National Nanotechnology Initiative, providing appropriations for the conduct of nanoscale science, engineering, and technology research; establishing programs; and creating a legislative foundation for the activities of the NNI.

Congressional funding for the NNI is provided through appropriations to each of the NNI-participating agencies. The NNI has no centralized funding. The overall NNI budget is calculated by aggregating the nanotechnology budgets for each of the federal agencies that conduct or provide funding for nanoscale science, engineering, and technology research.

In FY2001, the first year of NNI funding, Congress provided $464 million to eight agencies for nanoscale science, engineering, and technology research.[39] The NNI has continued to receive support from both Congress and the White House. Both the number of agencies participating in the NNI and the size of the federal investment have grown. Currently 25 agencies participate in the NNI, 13 of which have received appropriated funds for nanotechnology R&D in FY2009.[40] Total NNI funding in FY2009 is approximately $1.8 billion, more than three times the level of funding provided in FY2001. The original six agencies identified at the launch of the NNI[41] still account for the vast majority of NNI funding (97.3% in FY2009; detailed agency funding levels for FY2010 are not yet available).

21st Century Nanotechnology Research and Development Act of 2003

Congress codified and further defined some of the NNI's activities in the 21st Century Nanotechnology Research and Development Act of 2003 which was passed by Congress in November 2003, and signed into law (P.L. 108-153) by President Bush on December 3, 2003.[42] The legislation received strong bipartisan support in both the House of Representatives, which passed the bill on a recorded vote of 405-19, and in the Senate, which passed the bill by unanimous consent.

Though this act is often referred to as the enabling legislation for the National Nanotechnology Initiative, the act actually establishes a National Nanotechnology Program (NNP). The act provides authorizations for five NNI agencies—the National Science Foundation, Department of Energy, NASA, National Institute of Standards and Technology, and Environmental Protection Agency—but not for the Department of Defense, National Institutes of Health, Department of Homeland Security,[43] or other NNI research agencies that collectively accounted for 46% of NNI funding in FY2003.

The act created the NNP for the purposes of establishing the goals, priorities, and metrics for evaluation of federal nanotechnology research, development, and other activities; investing in federal R&D programs in nanotechnology and related sciences to achieve those goals; and providing for interagency coordination of federal nanotechnology research, development, and other activities undertaken pursuant to the NNP.

Key provisions of the act include:

- authorizing appropriations for the nanotechnology-related activities of the National Science Foundation, Department of Energy, NASA, National Institute of Standards and Technology, and Environmental Protection Agency for fiscal years 2005 through 2008, totaling $3.679 billion for the four year period;
- establishing a National Nanotechnology Coordination Office, with a director and full time staff to provide administrative support to the NSTC;
- establishing a National Nanotechnology Advisory Panel (NNAP) to advise the President and the NSTC on matters relating to the NNP.
- establishing a triennial review of the NNP by the National Research Council of the National Academies of Sciences;
- directing the NSTC to oversee the planning, management, and coordination of the program, including the development of a triennial strategic plan;
- directing the Department of Commerce's National Institute of Standards and Technology to establish a program to conduct basic research on issues related to the development and manufacture of nanotechnology, and to use the Manufacturing Extension Partnership program to ensure results reach small- and medium-sized manufacturing companies;
- directing the Secretary of Commerce to use the National Technical Information Service to establish a clearinghouse of information related to commercialization of nanotechnology research;
- directing the Secretary of Energy to establish a program to support consortia to conduct interdisciplinary nanotechnology R&D designed to integrate newly developed nanotechnology and microfluidic tools with systems biology and molecular imaging;
- directing the Secretary of Energy to carry out projects to develop, plan, construct, acquire, operate, or support special equipment, instrumentation, or facilities for investigators conducting nanotechnology R&D; and
- directing the establishment of two centers, on a merit-reviewed and competitive basis: (1) the American Nanotechnology Preparedness Center, to conduct, coordinate, collect, and disseminate studies on the societal, ethical, environmental, educational, legal, and workforce implications of nanotechnology; and to identify

anticipated issues related to the responsible research, development, and application of nanotechnology, as well as provide recommendations for preventing or addressing such issues, and (2) the Center for Nanomaterials Manufacturing, to encourage, conduct, coordinate, commission, collect, and disseminate research on new manufacturing technologies for materials, devices, and systems with new combinations of characteristics, such as, but not limited to, strength, toughness, density, conductivity, flame resistance, and membrane separation characteristics; and to develop mechanisms to transfer such manufacturing technologies to U.S. industries.

While the act establishes a National Nanotechnology Program, the Executive Branch continues its broader effort under the NNI framework and name. According to the NNI's 2004 Strategic Plan:

> Many of the activities outlined in the Act were already in progress as part of the NNI. Moreover, the ongoing management of the initiative involves considerable input from Federal agencies that are not named specifically in the Act.... For continuity, and to capture this broader participation, the coordinated Federal activities as a whole will continue to be referred to as the National Nanotechnology Initiative.[44]

Reauthorization Efforts

The 21st Century Nanotechnology Research and Development Act provided a legislative foundation for some of the activities of the NNI, authorized agency funding levels through FY2008, and sought to address challenges associated with the development and commercialization of nanotechnology. While many provisions of this act have no sunset provision, FY2008 was the last year of agency authorizations included in the act. Legislation to amend and reauthorize the act was introduced in the House (H.R. 5940, 110th Congress) and the Senate (S. 3274, 110th Congress) in the 110th Congress. Both bills were titled the National Nanotechnology Initiative Amendments Act of 2008. The House passed H.R. 5940 by a vote of 407-6; the Senate did not act on S. 3274.

In January 2009, H.R. 554, the National Nanotechnology Initiative Amendments Act of 2009, was introduced in the 111th Congress. The act contains essentially the same provisions as H.R. 5940 (110th Congress). (For additional information on H.R. 554, see bill highlights on page 38.) In February 2009, the House passed the bill by voice vote under a suspension of the rules. In July 2009, S. 1482, the National Nanotechnology Initiative Amendments Act of 2009, was introduced in the Senate. The bill was referred to the Senate Committee on Commerce, Science, and Transportation. No further action has been taken. H.R. 820, the Nanotechnology Advancement and New Opportunities Act, also would amend P.L. 108-153. The provisions of H.R. 820 cover a variety of jurisdictions, thus the bill has been assigned to multiple House committees. On May 7, 2010, the House Committee on Science and Technology reported the America COMPETES Reauthorization Act of 2010 (H.R. 5116) which includes, as Title I, Subtitle A, the "National Nanotechnology Initiative Amendments Act of 2010." Provisions of this subtitle are nearly identical to the provisions of H.R. 554. (For additional information on these bills, see "Selected Nanotechnology Legislation in the 111th Congress.")

Congress may use these bills to further address these issues and to establish authorization levels for agency nanotechnology R&D. Alternatively, Congress may choose to address some or all of these issues in separate legislation. Several bills have been introduced in the first session of the 111th Congress to address specific nanotechnology issues.

Structure

Nanoscale Science, Engineering, and Technology Subcommittee

The NNI is coordinated within the White House through the NSTC, the Cabinet-level council by which the President coordinates science, space, and technology policies across the federal government. Operationally, NNI coordination is accomplished through the Nanoscale Science, Engineering, and Technology (NSET) subcommittee of the NSTC's Committee on Technology (CT). The NSET subcommittee also has an informal reporting relationship to the NSTC's Committee on Science (CS). The NSET subcommittee is led by an agency co-chair, currently from the Department of Energy (DOE), and an OSTP co-chair. The NSET subcommittee is comprised of representatives from 25 federal entities (including 15 that fund nanotechnology R&D), OSTP and the Office of Management and Budget.[45]

The NSET subcommittee has established several working groups, each taking on efforts in key subject areas.[46] Among them:

National Environmental and Health Implications (NEHI)

The National Environmental and Health Implications (NEHI) working group was chartered to provide for exchange of information among agencies that support research and those responsible for regulations and guidelines related to nanotechnology products; to facilitate identification, prioritization, and implementation of research and other activities required for the responsible research, development, utilization, and oversight of nanotechnology; and to promote communication of information related to research on environmental and health implications of nanotechnology to other government agencies and non-government parties. To this end, the NEHI working group seeks to identify and prioritize environmental, health, and safety research needs related to nanotechnology. Twenty NNI agencies participate in the NEHI working group, and 13 agencies fund safety-related nanotechnology research and/or have regulatory authorities to guide the safe use of nanomaterials.[47]

Nanomanufacturing, Industry Liaison, and Innovation(NILI)

The Nanomanufacturing, Industry Liaison, and Innovation (NILI) working group was chartered to enhance collaboration and information sharing between U.S. industry and government on nanotechnology-related activities. It also facilitates federal, regional, state, and local nanotechnology R&D and commercialization activities. In addition, the NILI working group is to create innovative methods for transferring federally funded technology to industry. The NILI working group has facilitated collaborations between the NNI and the semiconductor/electronics industry, chemical industry, forest products industry, and the Industrial Research Institute.[48]

Global Issues in Nanotechnology (GIN)

The Global Issues in Nanotechnology (GIN) working group was chartered to monitor foreign nanotechnology programs and development; broaden international collaboration on nanotechnology R&D, including safeguarding the environment and human health; and promote U.S. commercial and trade interests in nanotechnology. The NEHI working group works with the GIN working group to coordinate the U.S. position and participation in international activities related to environmental, health, and safety implications of nanotechnology. The GIN working group facilitates international collaboration on pre-competitive and non-competitive aspects of nanotechnology, and international engagement on trade, commercialization and regulatory issues.

Nanotechnology Public Engagement and Communications (NPEC)

The Nanotechnology Public Engagement and Communications (NPEC) working group was established to develop approaches by which the NNI can communicate more effectively with the public.

National Nanotechnology Coordination Office

The National Nanotechnology Coordination Office (NNCO) provides administrative and technical support to the NSET subcommittee. Initially established in 2001 through a memorandum of understanding among the NNI participating agencies,[49] the NNCO was authorized by the 21st Century Nanotechnology Research and Development Act of 2003 (P.L. 108-153). The NNCO was charged under the act with providing technical and administrative support to the NSTC and NNAP; serving as the point of contact for information on Federal nanotechnology activities for the exchange of technical and programmatic information among stakeholders; conducting public outreach; and promoting access to and early application of NNP technologies, innovation, and expertise.

The act authorizes the work of the NNCO to be funded by contributions from NSET subcommittee member agencies. According to the NNCO, funding is provided through a memorandum of understanding signed by eight NNI agencies.[50] In principle, each agency contributes to the NNCO budget in proportion to its share of the President's total nanotechnology budget request for the signatory agencies. However, two of the signatories, EPA and DOT, had sufficiently small enough nanotechnology budgets in the early years of the NNI that they were not expected to contribute. EPA now contributes to funding the NNCO. Total NNCO funding from the agencies in FY2010 is $2.4 million. In addition, the NNCO has carried over FY2009 funds intended to support a National Academies' study, bringing the total NNCO budget for FY2010 to just under $3 million.

Funding

The NNI supports fundamental and applied research on nanotechnology by funding research, creating multidisciplinary centers of excellence, and developing key research infrastructure. It also supports activities aimed at addressing the societal implications of nanotechnology, including ethical, legal, human and environmental health, and workforce issues.

This section provides information on NNI funding from two perspectives: organizationally by agency and functionally by program component area.

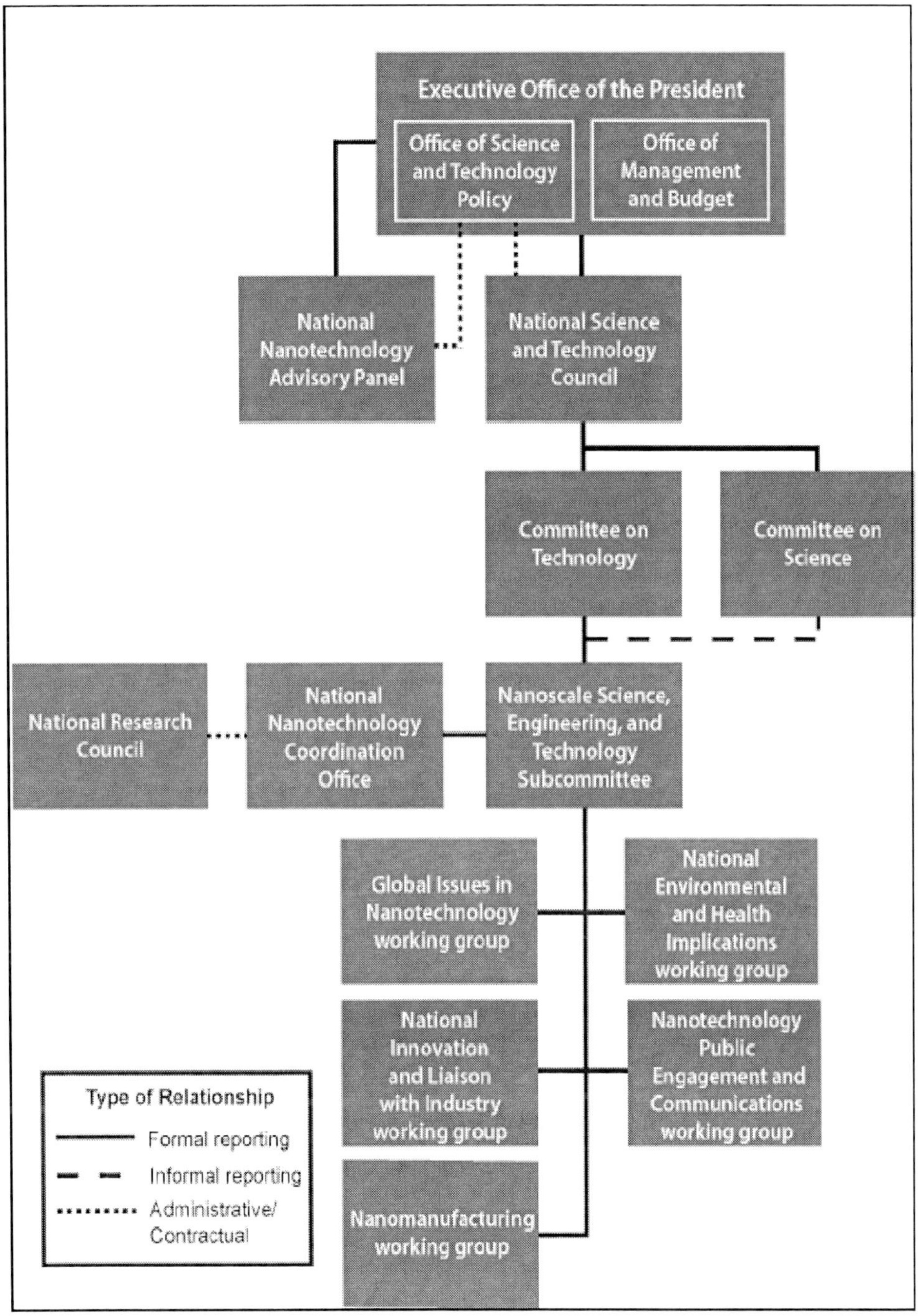

Source: *The National Nanotechnology Strategic Plan*, Nanoscale, Science, Engineering, and Technology Subcommittee, National Science and Technology Council, The White House, December 2007. (Nanomanufacturing working group added to chart in cited source.)

Figure 1. Organizations With a Role in the National Nanotechnology Initiative and Their Relationships

Table 1. NNI Funding, by Agency: FY2001-FY2009 and FY2010 Request (in millions of current dollars)

Agency	FY 2001 Actual	FY 2002 Actual	FY 2003 Actual	FY 2004 Actual	FY 2005 Actual	FY 2006 Actual	FY 2007 Actual	FY 2008 Actual	FY 2009 Actual	ARRA (P.L. 111-5)[a]	FY2010 Estimate	FY2011 Request
National Science Foundation	150	204	221	256	335	360	389	409	409	101	418	401
Department of Defense[b]	125	224	220	291	352	424	450	460	459		436	349
Department of Energy	88	89	134	202	208	231	236	245	333	293	373	424
National Institutes of Health (HHS)	40	59	78	106	165	192	215	305	343	73	361	382
Nat'l Inst. of Standards and Tech. (DOC)	33	77	64	77	79	78	88	86	93	43	114	108
Environmental Protection Agency	5	6	5	5	7	5	8	12	12		18	20
NASA	22	35	36	47	45	50	20	17	14		14	16
Department of Justice	1	1	1	2	2	<1	2	0	1		0	
Department of Homeland Security		2	1	1	1	2	2	3	9		12	12
Nat'l Inst. of Food and Agriculture (USDA)[c]			1	2	3	4	4	6	10		10	9
Nat'l Inst. for Occupational Safety (HHS)					3	4	7	7	7		10	17
Forest Service (USDA)						2	3	5	5		5	5
Federal Highway Administration (DOT)						1	1	1	1		3	2
Food and Drug Administration (HHS)									7		7	15
Consumer Product Safety Commission									<1		<1	2
TOTAL[d]	464	697	760	989	1,200	1,351	1,425	1,554	1,702	511	1,781	1,762

Sources: NNI website, http://www.nano.gov.

a. Funding figures for nanotechnology-related R&D under the ARRA are preliminary estimates.

b. According to NSTC, the DOD budgets shown above for FY2006, FY2007, FY2008, and FY2009 include congressionally directed funding of approximately $76 million in FY2006, $63 million in FY2007, and $117 million in FY2009; the NSTC states that the DOD budget for FY2010 also includes congressionally directed funding but does not provide an amount. According to NSTC, the 2008 DOD estimate "includes many earmarks that are outside the NNI plan."

c. Formerly, the USDA Cooperative State Research, Education, and Extension Service (CSREES).

d. Totals may not add due to rounding of agency budget figures.

Table 2. NNI Funding, by Program Component Area, FY2006-FY2009 (in millions of current dollars)

PCA	FY2006 Actual	FY2007 Actual	FY2008 Actual	FY2009 Actual	ARRA Preliminary	FY2010 Estimate	FY2011 Request
Fundamental Phenomena and Processes	455.9	480.6	478.5	479.2	130.6	467.9	484.4
Nanomaterials	265.1	258.3	285.1	331.9	178.3	373.5	342.3
Nanoscale Devices and Systems	319.6	344.7	372.7	435.2	68.0	429.9	402.0
Instrumentation Research, Metrology, and Standards	51.0	52.5	69.0	90.8	12.4	84.3	76.9
Nanomanufacturing	33.8	48.1	47.1	75.6	28.5	96.7	101.4
Major Research Facilities and Instrumentation Acquisition	152.4	152.4	196.4	177.6	72.5	197.3	203.0
Societal Dimensions	73.5						
• Environment, Health, and Safety		48.3	67.9	74.5	12.0	91.6	116.9
• Education & Societal Dimensions		39.2	37.7	36.8	9	39.9	34.8
TOTAL[a]	**1,351.2**	**1,424.1**	**1,554.4**	**1,701.5**	**511.3**	**1,781.1**	**1,761.6**

Source: *The National Nanotechnology Initiative: Research and Development Leading to a Revolution in Technology and Industry, Supplement to the President's FY2008 Budget*, Nanoscale Science, Engineering, and Technology Subcommittee, National Science and Technology Council, The White House, July 2007. National Nanotechnology Initiative: FY2009 Budget & Highlights, National Science and Technology Council, The White House, February 2008. *The National Nanotechnology Initiative: Research and Development Leading to a Revolution in Technology and Industry, Supplement to the President's FY2010 Budget*, Nanoscale Science, Engineering, and Technology Subcommittee, National Science and Technology Council, The White House, May 2009.

a. Details may not add to totals due to rounding.

Agency Funding

The NNI budget is an aggregation of the nanotechnology components of the individual budgets of NNI-participating agencies. The NNI budget is not a single, centralized source of funds that is allocated to individual agencies. In fact, agency nanotechnology budgets are developed internally as part of each agency's overall budget development process. These budgets are subjected to review, revision, and approval by the Office of Management and Budget and become part of the President's annual budget submission to Congress. The NNI budget is then calculated by aggregating the nanotechnology components of the appropriations provided by Congress to each federal agency.

For FY2010, regular appropriations for NNI agencies totaled an estimated $1.781 billion,[51] a $79.5 million (4.7%) increase over the FY2009 regular appropriation and nearly four times the $464 million federal investment in nanotechnology research in FY2001. This growth in nanotechnology R&D investments reflects expectations in Congress and in the executive branch that the NNI will expand fundamental knowledge and make important contributions to national priorities. The chronology of NNI funding is detailed in **Table 1**.

President Obama has requested $1.762 billion in funding for the NNI in FY2011, a decrease of $19.5 million (1.1%). The FY2011 NNI budget request would support a broad range of programs among 13 agencies. Agencies with the largest budgets are:

- NSF, which supports fundamental nanotechnology research across science and engineering disciplines;
- DOD, whose investments in nanotechnology are aimed at addressing the department's national security mission;
- DOE, which supports nanotechnology research providing a basis for new and improved energy efficiency, production, storage, and transmission technologies;
- NIH, which emphasizes nanotechnology-based biomedical advances occurring at the intersection of biology and the physical sciences; and
- NIST, which focuses on research in instrumentation, measurement, standards, characterization, and nanomanufacturing.

Other agencies investing in mission-related nanotechnology R&D are NASA, EPA, the Cooperative State Research, Education, and Extension Service (CSREES) and Forest Service at the Department of Agriculture (USDA), National Institute of Occupational Safety and Health (NIOSH), DHS, Department of Justice (DOJ), and Department of Transportation's (DOT's) Federal Highway Administration (FHWA).

Program Component Area Funding

The 21st Century Nanotechnology R&D Act of 2003 called for the NSET subcommittee to develop categories of investment called Program Component Areas (PCA) to provide a means by which Congress and the executive branch can be informed of and direct the relative investments in these areas. The PCAs are categories of investments that cut across the needs and interests of individual agencies and contribute to the achievement of one or more of the NNI's goals. The 2004 NNI strategic plan identified seven PCAs. The 2007 NNI strategic plan splits the seventh PCA, Societal Dimensions, into two PCAs: Environment, Health, and Safety; and Education and Societal Dimensions. A description of the seven initial PCAs and their current funding are provided below,[52] as well as a description of the two derivative PCAs.[53] The chronology of NNI funding by PCA is detailed in **Table 2**.

In the following analysis of funding for each of the PCAs, comparisons involving FY2009 funding (i.e., between FY2010 and FY2009) exclude ARRA funding. In addition, comparisons between the President's FY2011 budget request and estimated FY2010 appropriations may be affected by the President's exclusion of congressionally directed funding in his budget.

Fundamental Phenomena and Processes

Fundamental Phenomena and Processes includes investments in the discovery and development of fundamental knowledge pertaining to the new phenomena in the physical, biological, and engineering sciences that occur at the nanoscale, as well as in understanding and articulation of scientific and engineering principles related to nanoscale structures, processes, and mechanisms.

FY2010 funding for Fundamental Phenomena and Processes fell to $467.9 million, down $11.3 million (2.4%) from the FY2009 level due primarily to a decrease in DOD funding in this PCA (down $24.0 million, 14.7%). President Obama's FY2011 budget proposes $484.4 million in funding, up $16.5 million (3.5%) over the FY2010 level due primarily to increases in DOE ($14.2 million, 13.8%) and DOD ($412.7 million, 9.1%) funding. The increase in DOE and DOD funding would somewhat offset by a decrease in NSF funding of $12.5 million (8.2%).[54]

Nanomaterials

Nanomaterials includes research investments to discover novel nanoscale and nanostructured materials. This PCA also attempts to understand the properties of nanomaterials, and supports R&D to enable the design and synthesis, in a controlled manner, of nanoscale materials with targeted properties.

FY2010 funding for Nanomaterials rose to $373.5 million, up $41.6 million (12.5%) from the FY2009 level, led by an increase in DOE funding in this PCA (up $21.7 million, 23.4%). President Obama's FY2011 budget proposes $342.3 million for this PCA, a decrease of $31.2 million (8.4%) from the FY2010 level, resulting primarily from reduced DOD funding (down $36.0 million, 47.8%).[55]

Nanoscale Devices and Systems

Nanoscale Devices and Systems include R&D investments that apply nanoscale science and engineering principles to create novel devices and systems or to improve existing ones. It also includes the use of nanoscale or nanostructured materials to achieve improved performance or new functionality. To meet this definition, the enabling science and technology must be at the nanoscale, but the systems and devices are not restricted to that size.

Funding for Nanoscale Devices and Systems fell to $429.9 million in FY2010, down $5.3 million (1.2%) from the FY2009 level. President Obama's FY2011 budget proposes $402.0 million in funding for this PCA, a decrease of $27.9 million (6.5%) from the FY2010 level, largely due to reductions in DOD funding (down $49.0 million, 33.1%). The decrease in

DOD funding for this PCA would somewhat offset by increases in DOE ($13.2 million, 76.7%) and NIH funding ($13.1 million, 7.2%).[56]

Instrumentation Research, Metrology, and Standards

The Instrumentation Research, Metrology, and Standards PCA includes R&D investments for development of tools needed to advance nanotechnology research and commercialization. Instrumentation for characterization, measurement, synthesis, and design of nanotechnology materials, structures, devices, and systems is funded through this PCA. R&D and other activities related to development of standards, including standards for nomenclature, materials, characterization, testing, and manufacture are also in this PCA.

FY2010 funding for Instrumentation Research, Metrology, and Standards fell to $84.3 million, down $6.5 million (7.2%) from the FY2009 level. President Obama's FY2011 budget proposes $76.9 million in funding for this PCA, a decrease of $7.4 million (8.8%) from the FY2010 level, due to decreases in DOD, DOE, and NSF funding.[57]

Nanomanufacturing

Nanomanufacturing R&D supports the development of scalable, reliable, cost-effective manufacturing of nanoscale materials, structures, devices, and systems. It also includes R&D and integration of ultra-miniaturized top-down processes and complex bottom-up processes.[58]

FY2010 funding for Nanomanufacturing rose to $96.7 million, up $21.1 million (27.9%) over the FY2009 level due primarily to increases in NIST ($17.8 million, 189.4%) and DOD ($8.2 million, 28.3%) funding. President Obama's FY2011 budget proposes $101.4 million for this PCA, an increase of $4.7 million (4.9%) above the FY2010 level as increases in DOE ($13.9 million, 198.6%) and NSF ($9.8 million, 43.8%) would more than offset decreases in DOD ($12.1 million, 32.5%) and NIST ($7.0 million, 25.7%) funding.[59]

Major Research Facilities and Instrumentation Acquisition

This PCA includes investments in the establishment and ongoing operations of user facilities and networks, the acquisition of major instrumentation, and other activities related to infrastructure for the conduct of nanoscale science, engineering, and technology R&D.

FY2010 funding for Major Research Facilities and Instrumentation Acquisition rose to $197.3 million, an increase of $19.7 million (11.1%) over the FY2009 level. President Obama's FY2011 budget proposes $203.0 million for this PCA, an increase of $5.7 million (2.9%) above the FY2009 level, led by increases in DOE and DOD funding.[60]

Societal Dimensions

The Societal Dimensions PCA includes investments in research and other activities that address the broad implications of nanotechnology to society. This includes assessing benefits and risks through research directed at environmental, health, and safety impacts of nanotechnology development; risk assessment of such impacts; education-related activities, such as development of materials for schools, undergraduate programs, technical training, and public outreach; and research directed at identifying and quantifying the broad implications of nanotechnology for society, including social, economic, workforce, educational, ethical, and legal implications.

Under the 2007 NNI Strategic Plan, the Societal Dimensions PCA was divided into two separate PCAs: Environment, Health, and Safety, and Education and Societal Dimensions. Future PCA reporting will use the new eight PCA taxonomy. NSTC has retroactively reported FY2007 Societal Dimensions PCA spending in the new PCAs.[61] The NSET subcommittee characterizes the new PCAs as follows:[62]

Environment, Health, and Safety

This PCA addresses research primarily directed at understanding the environmental, health, and safety impacts of nanotechnology development and corresponding risk assessment, risk management, and methods for risk mitigation.

FY2010 funding for Environment, Health, and Safety rose to $91.6 million, up $17.1 million (23.0%) above the FY2009 level, due to increases in EPA, NIH, NSF, and NIOSH funding. President Obama's FY2011 budget proposes $116.9 million in funding for this PCA, an increase of $25.3 million (27.6%) above the FY2010 level, led by increases at FDA and NIOSH.[63]

Education and Societal Dimensions

This PCA addresses education-related activities such as development of materials for schools, undergraduate programs, technical training, and public communication, including outreach and engagement. Such activities include research directed at identifying and quantifying the broad implications of nanotechnology society, including social, economic, workforce, educational, ethical, and legal implications.

FY2010 funding for Education and Societal Dimensions rose slightly to $39.9 million, up $3.1 million (8.4%) from the FY2009 level due primarily to an increase in NSF funding ($3.0 million, 9.6%); NSF accounts for 86.0% of funding in this PCA in FY2009. President Obama's FY2011 budget proposes $34.8 million in funding for this PCA, a decrease of $5.1 million (12.8%) from the FY2010 level, accounted for entirely by a decrease in NSF funding.[64]

Centers, Networks, and User Facilities

A key facet of the National Nanotechnology Initiative has been the development of an extensive infrastructure of interdisciplinary research and education centers, networks, and user facilities. The centers and user facilities are located at universities and federal laboratories across the country.

Centers and networks provide opportunities and support for multidisciplinary research among investigators from a variety of disciplines and research sectors, including academia, industry, and government laboratories. Such multidisciplinary research not only can lead to advances in knowledge, but also may foster relationships that further the development of basic research results into devices and other applications.

Many agencies support such centers. Examples of federal and federally supported centers include:

- The National Science Foundation has established university-based centers focused exclusively on nanotechnology, including 15 Nanoscale Science and Engineering Centers (NSEC), one Engineering Research Center, one Science and Technology Center, four Materials Research Science and Engineering Centers, two Nanoscale Science and Engineering Education Centers, and five Nanoscale Science and Engineering Networks.[65]
- The NIH has established more than 20 centers, including eight university-based Nanomedicine Development Centers; a Nanotechnology Characterization Laboratory, established by the National Cancer Institute (NCI), in partnership with NIST and the Food and Drug Administration; eight university-based Centers of Cancer Nanotechnology Excellence, established under the NCI's Alliance for Nanotechnology in Cancer initiative; and four university-based centers, established by the National Heart, Lung, and Blood Institute under its Program of Excellence in Nanotechnology.
- The Department of Defense supports two university-based nanotechnology research centers, as well as the Institute for Nanoscience at the Naval Research Laboratory.
- NASA has established three centers under its University Research, Engineering, and Technology Institute program.
- The Department of Energy has established five Nanoscale Science Research Centers (NSRCs) co-located with its national labs.
- NIST has established a Center for Nanoscale Science and Technology (CNST).
- NIOSH has established a Nanotechnology Research Center to conduct research into the application of nanoparticles and nanomaterials in occupational safety and health and the implications of nanoparticles and nanomaterials for work-related injury and illness.

Many of the centers are designated as user facilities and are available to researchers not located at the center. User facilities are designed to allow outside researchers to take advantage of facilities, equipment, tools, and expertise. These shared resources provide researchers the opportunity to conduct research, characterize materials, and test products using equipment and facilities that their individual companies, universities, or organizations could not afford to acquire, support, or maintain. Conditions for user access vary by facility and agency. In general, users are not charged for pre-competitive, non-proprietary work leading to publication, and are charged on a cost-recovery basis for proprietary work. In some cases, the user facilities are located at federal government laboratories (e.g. the Department of Energy's five NSRCs, and the NIST CNST); other user facilities are located at universities and supported with federal funds (e.g. NSF's 13 university-based centers in the National Nanotechnology Infrastructure Network (NNIN)).

As mentioned earlier, the 21st Century Nanotechnology R&D Act of 2003 directed the establishment of two centers, the American Nanotechnology Preparedness Center and the Center for Nanomaterials Manufacturing. According to the NSET subcommittee, the requirement to establish the American Nanotechnology Preparedness Center was met by NSF's establishment of the Network for Nanotechnology in Society, comprised of centers at the University of California, Santa Barbara (with the participation of Harvard University and the University of South Carolina) and the University of Arizona.[66] These centers were funded under NSF's Nanoscale Science and Engineering Center (NSEC) program and did not include participation by any other NSET subcommittee agency.[67] The NSET subcommittee states that the requirement for establishing the Center for Nanomaterials Manufacturing was met by NSF's establishment of a National Nanomanufacturing Network (NNN) comprised of four NSECs. The Center for Integrated Hierarchical Manufacturing at the University of Massachusetts Amherst is the main node of the NNN.[68] The NNN NSECs were established by NSF in collaboration with DOD and NIST, but exclusively with NSF funds.[69]

SELECTED NNI REPORTS AND ASSESSMENTS

This section presents summaries of recent reports from the NSTC's Nanoscale Science, Engineering, and Technology Subcommittee and assessments conducted by the National Research Council and the President's Council of Advisors on Science and Technology.

Selected NNI Reports

The NNI's coordinating body, the NSTC's Nanoscale Science, Engineering, and Technology Subcommittee, produces a variety of reports that serve to inform Congress and other key stakeholders on the initiatives' current activities, investments, and priorities.

The National Nanotechnology Initiative: Research and Development Leading to a Revolution in Technology and Industry, Supplement to the President's FY2011 Budget[70]

Each year the NSET subcommittee publishes a supplement to the President's annual budget request. The FY2011 NNI budget supplement provides a more detailed look at NNI funding in the President's budget request, including a break-out of the prior, current, and requested year budgets for each PCA. The report describes proposed changes in agency R&D budgets, as well as in the balance of investments by PCA. Of particular note:

- President Obama has proposed an overall NNI budget for FY2011 of $1.762 billion, a $19.5 million (1.1%) decrease below FY2010 funding.
- Estimated funding for EHS R&D in FY2010 rose to $91.6 million, a 23.0% increase over FY2009. President Obama has requested $116.9 million for EHS R&D in FY2011, an increase of $25.3 million (27.6%) over FY2010.
- President Obama's FY2011 budget request reflected a decline in spending in several program component areas compared to FY2010 estimated funding levels, including nanomaterials, down $12.0 million (3.9%); nanoscale devices and systems, down $31.2 million (8.4%); nanoscale devices and systems, down $27.9 million (6.5%), and instrumentation research, metrology and standards, down $7.4 million (8.8%). In addition to EHS funding, other program component areas that would be increased include fundamental phenomena and processes, up $16.5 million (3.5%), major research facilities and instrumentation acquisition, up $5.7 million (2.9%), and nanomanufacturing, up $4.7 million (4.9%%).

The National Nanotechnology Strategic Plan (2007)[71]

The 21st Century Nanotechnology R&D Act of 2003 (P.L. 108-153) requires the NSTC to develop an NNI strategic plan every three years. This plan is to guide the program's activities to meet the goals, priorities, and anticipated outcomes of the participating agencies. In addition, the act requires the triennial report to address how the program intends to move results out of the laboratory and into application for the benefit of society, its plan for long-term funding for interdisciplinary R&D, and the allocation of funding for interagency projects. The 2007 strategic plan is the first to follow external assessments by the National Academies and PCAST (operating as the NNAP) and seeks to incorporate the findings of these reviews. Of particular note:

- The 2007 strategic plan includes a new chapter on "High-Impact Application Opportunities and Critical Research Needs" possibly indicating an effort on the part of the Administration to move the NNI toward more directed research with commercial and societal benefits. Much of the early NNI work has been focused on basic research and mechanisms by which such research may produce economic and societal dividends. Seven years into the NNI, the program is under increasing scrutiny to deliver the promised benefits. While the plan does not establish R&D or application targets per se, this chapter illustrates tangible benefits that may be achieved by research supported under the NNI. In addition, the plan provides a matrix that identifies which agencies have a central or supporting role in each key application area. The plan also provides a series of nanotechnology application-specific vignettes on topics such as early detection of life-threatening disease, smarter computers, more energy-efficient transportation, and energy security.
- The 2007 strategic plansplit the Societal Dimensions PCA into two separate PCAs: Environmental, Health, and Safety, and Education and Societal Dimensions. This change responds to increased Congressional and public attention to EHS needs. Some critics of the NNI had raised concerns that the inclusion of investments in education and other societal dimensions in the broader category obscured and artificially inflated the perception of investments in EHS R&D.
- The 2007 strategic planalso identifies four areas of common interest across agencies that is to be the focus of future workshops: sensors and nanoelectronics, energy, fate and transport of nanomaterials, and medical and health applications.

Strategy for Nanotechnology-related Environmental, Health, and Safety Research[72]

Strategy for Nanotechnology-related Environmental, Health, and Safety Research outlines the NNI's strategy for addressing nanotechnology EHS concerns. The report outlines the process for developing a research strategy, including identification of priority needs, assessment of existing research, and analysis of strengths and weaknesses of the current portfolio of EHS-focused research. The report provides a summary and analysis of FY2006 EHS research projects using the five priority categories identified in the earlier report, *The National Nanotechnology Initiative: Environmental, Health, and Safety Research Needs for Engineered Nanoscale Materials*[73] (discussed below). The report concludes with a framework for addressing EHS research needs, including a summary and prioritization of research needs, an implementation strategy, and a discussion of interagency coordination efforts.

***Prioritization of Environmental, Health, and Safety Research Needs for Engineered Nanoscale Materials: An Interim Document for Public Comment*[74]**

This document is a follow-on to the *Environmental, Health, and Safety Research Needs for Engineered Nanoscale Materials* report (discussed below), incorporating public comments, refinements of the prioritization principles, and continued assessment of research needs. This report further identifies and defines five priorities within each of the five general categories of research needs established in the earlier document and presents the revised principles and the process used for this prioritization. The NEHI working group of the NSET subcommittee expects to use this report to evaluate the NNI's current EHS research portfolio, perform a gap analysis, and identify opportunities for interagency collaboration. The report stresses that the NSET subcommittee is "pursuing a dynamic, open, and transparent process in developing an NNI EHS research strategy" and invites continuing public input.

***The National Nanotechnology Initiative: Environmental, Health, and Safety Research Needs for Engineered Nanoscale Materials*[75]**

This report documents the efforts of the NSET subcommittee's Nanotechnology Environmental and Health Implications (NEHI) working group to identify, prioritize, and implement research and other activities required for the responsible research and development of nanotechnology. The report is designed to help inform the research, risk assessment, and risk management activities of federal agencies and the private sector.

The report identifies priority research within five general research areas: instrumentation, metrology, and analytical methods; nanomaterials and human health; nanomaterials and the environment; health and environmental surveillance; and risk management methods.

The report identifies several next steps :

- prioritize research needs among those identified in the report;
- evaluate in greater detail the current NNI EHS research portfolio;
- perform a "gap analysis" of the NNI EHS research compared to the prioritized needs;
- coordinate and facilitate among the NNI agencies' research programs to address priorities; and
- establish a process for periodic review of progress and for updating research needs and priorities.

The report concludes that conducting EHS research in parallel with the development of nanomaterials and their applications will help to ensure the full, safe, and responsible realization of the promise of nanotechnology, and that coordination of research activities among NNI-participating agencies, as well as with industry and other governments, is necessary to expedite progress. In contrast, some NGOs have asserted the need for EHS research to precede the development of nanomaterials and nanotechnology applications.

Selected NNI Assessments

The 21st Century Nanotechnology R&D Act of 2003 (P.L. 108-153) requires periodic external reviews of the National Nanotechnology Program (NNP) by the National Research

Council, an arm of the National Academies,[76] and the National Nanotechnology Advisory Panel (NNAP).[77] In general, these reviews concluded that the NNI has been successful so far and that its efforts are important to future U.S. technological leadership and commercial competitiveness. Both reports emphasize that much nanotechnology research is still in its very early stages and caution against expecting too much in the near term from this nascent technology. The reports also laud the cooperative efforts between the NNI and stakeholders in academia and industry and encourage increased interactions with industry, state and local economic developers, and, where appropriate, international partners.

The National Nanotechnology Initiative: Second Assessment and Recommendations of the National Nanotechnology Advisory Panel, President's Council of Advisors on Science and Technology[78]

This report presents the findings of the second review of the NNI by the President's Council of Advisors on Science and Technology's, acting as the National Nanotechnology Advisory Panel, as mandated by the 21st Century Nanotechnology R&D Act of 2003. PCAST submitted its first report to the President on May 16, 2005, titled *The National Nanotechnology Initiative at Five Years: Assessment and Recommendations of the National Nanotechnology Advisory Panel.*

As in earlier reports by PCAST and the National Research Council, this report concludes that the United States remains a leader in nanotechnology, though it recognizes the successful development of other nations' nanotechnology research capabilities. The PCAST concluded that the NNI's approach to addressing EHS research was "sound" and found proposals for a separate agency or office devoted to nanotechnology EHS research or to set aside a particular percentage of NNI funding for EHS research to be "misguided" and potentially counterproductive as resources may be directed away from research "on beneficial applications and on risk." The panel also concluded that nanotechnology does not raise ethical concerns unique from those accompanying other technological advances. The PCAST also concluded that the NNI is a "highly successful model for an interagency program" describing it as "well organized and well managed."

Recommendations

The report made six broad recommendations for further strengthening the NNI, supported by more specific actions. They are:

PCAST recommendation:

> Maintaining the world-class R&D infrastructure and strong interagency coordination created under the NNI is essential to achieving broad societal benefits from nanotechnology innovation.

Some specific PCAST recommendations include continuing support for NNI multidisciplinary centers, networks and user facilities; improving coordination within agencies that have several operating divisions; and strengthening the participation of the Departments of Commerce, Education, and Labor to address education, training, market assessment, and standards development challenges related to nanotechnology.

PCAST recommendation:

> Progress across the breadth of NNI-supported R&D critically depends upon the development and implementation of standards for nanomaterial identification, characterization, and risk assessment.

In this regard, PCAST specifically recommended that federal agencies participate in the development of voluntary consensus-based standards; develop materials and analytical standards for nanotechnology EHS research; and work towards development of data sets of physical and chemical properties of nanomaterials.

PCAST recommendation:

> Nanotechnology innovation through to commercialization depends on maintaining and strengthening cross-sector collaborations and cross-fertilization of technology development and business development expertise.

Specifically, PCAST recommended expansion of efforts to assess national and international innovation and commercialization activities led by the Department of Commerce; expanded partnering between NNI multidisciplinary centers and economic development organizations; and educated more nanotechnology scientists and engineers to become entrepreneurs and skilled technology workers.

PCAST recommendation:

> Nanotechnology research must be strategically guided, integrated, and coordinated across agencies, sectors, and countries, and include balanced assessment of risks and benefits in the context of specific, real-world applications.

Among the specific PCAST recommendations: improving coordination of federal EHS research efforts with industry and international stakeholders; preventing the segregation of research focused on EHS implications of nanotechnology from research focused on applications; continuing interagency collaboration on EHS research and mission-focused research; and distributing nonproprietary information about the properties of nanomaterials.

PCAST recommendation:

> Research on the societal and ethical aspects of nanotechnology should both be integrated with technical R&D and take place in the context of broader societal and ethical leadership.

PCAST recommendation:

> Public perception of and expectations related to nanotechnology should be informed based on sound science and balanced assessment of risks and benefits (known and anticipated) of specific innovations and their implications for society.

PCAST recommended that the NNI more clearly demonstrate to the public the value of nanotechnology and NNI-supported R&D, calling for a more explicit and direct outreach approach to better inform and engage policymakers, stakeholders, and the general public. In addition PCAST recommended that NNI agencies should provided additional funding to the

NNCO to support coordination among NNI agencies to enhance their agency-specific communication efforts.

***Addendum to the National Nanotechnology Initiative: Second Assessment and Recommendations of the National Nanotechnology Advisory Panel, President's Council of Advisors on Science and Technology*[79]**

PCAST submitted an addendum to its second assessment of the NNI in July 2008, providing an assessment of *The National Nanotechnology Initiative Strategy for Nanotechnology-Related Environmental, Health, and Safety Research.* The PCAST letter reaffirmed its judgment (as stated in its second assessment of the NNI) of the NNI's approach to EHS research as "fundamentally sound in terms of strategic priorities as well as focus on and extent of collaboration across agencies, with industry, and particularly with international stakeholders." PCAST found the NNI's strategic approach to be "quite thorough and robust," and stated that the NSET subcommittee's NEHI working group

> remains the best locus of coordination and authoritative advisory capability for the participating agencies in implementing the government-wide nanotechnology EHS research strategy that cuts across agency and disciplinary lines.[80]

PCAST made six recommendations with respect to the NNI's EHS research strategy. First, PCAST recommended that an assessment of the federal nanotechnology EHS portfolio and gap analysis be performed once every three years noting that such a timetable would support the triennial reviews of PCAST and the NRC. Second, it proposed increased collaboration with industry and other countries to fill identified research gaps. Third, PCAST recommended that the NNI encourage its supported researchers to report on analytical methodologies used in their research to facilitate development of best practices for risk assessment and characterization. Fourth, PCAST stated that the NNI should promote broad and practical use of EHS findings in defining responsible use of nanotechnology in research, manufacturing, and commercial applications. Fifth, PCAST called for increased funding for exposure assessment in the context of manufacturing and disposal of nanomaterials and products incorporating relevant quantities of nanomaterials. Sixth, PCAST recommended that the NNI maintain and strengthen agency support and coordination efforts through the NSET subcommittee and the NEHI working group, specifically calling for

> all agencies that fund or conduct research on nanomaterials, as well as those agencies with charters that specifically include EHS-related activities, to directly support the NSET/NEHI data gathering and communications functions.

***A Matter of Size: Triennial Review of the National Nanotechnology Initiative*[81]**

This 2006 report presents the findings of the National Research Council's (NRC) first triennial review mandated by the 21st Century Nanotechnology R&D Act of 2003.

The NRC study concluded that the NNI has been successful in coordinating nanoscale efforts and interests across the federal government, in catalyzing cooperative R&D across a variety of scientific and engineering disciplines, and in opening a host of new scientific opportunities through its infrastructure and R&D investments. The NRC attributed much of this success to effective communication and coordination by the NSET subcommittee and the NNCO.

Recommendations

Here are the recommendations made by the NRC followed by a discussion of each.

NRC recommendation:

> the federal government [should] sustain investments in a manner that balances the pursuit of shorter-term goals with support for longer-term R&D and that ensures a robust supporting infrastructure, broadly defined. Supporting long-term research effectively will require making new funds available that do not come at the expense of much-needed ongoing investment in U.S. physical sciences and engineering research.

President Bush has expressed support for increasing federal R&D funding for the physical sciences and engineering, most notably in his American Competitiveness Initiative which includes nanotechnology investments. Yet, as Federal non-discretionary spending growth increases pressure on federal discretionary spending, finding new funds to support long-term nanotechnology research may need to come from other scientific disciplines.

NRC recommendation:

> the federal government [should] establish an independent advisory panel with specific operational expertise in nanoscale science and engineering; management of research centers, facilities, and partnerships; and interdisciplinary collaboration to facilitate cutting-edge research on and effective and responsible development of nanotechnology.

In July 2004, President George W. Bush implemented the provision of the 21st Century Nanotechnology Research and Development Act to "establish or designate a National Nanotechnology Advisory Panel" by issuing Executive Order 13349, which amends Executive Order 13226, designating the President's Council of Advisors on Science and Technology to serve as the NNAP.

The NRC's recommendation suggests that the President's designation of PCAST to serve as the legislatively mandated National Nanotechnology Advisory Panel is not fully adequate. Critics of the use of PCAST to serve as the NNAP maintain that the scope and depth of expertise needed to provide effective guidance on the NNI requires an independent panel of people with nanotechnology- and interdisciplinary-specific expertise and an undivided focus. Supporters of the use of PCAST for this function assert that a single advisory panel provides an integrated perspective, reduces unnecessary cost and management burdens, and that expertise can be added to the panel or accessed through non-member technical advisory groups.

NRC recommendation:

> federal agencies participating in the NNI, in consultation with the NNCO and the Office of Management and Budget, should continue to develop and enhance means for consistent tracking and reporting of funds requested, authorized, and expended annually. The current set of PCAs provides an appropriate initial template for such tracking.

It is difficult to assess and track funding for specific purposes within the NNI because the initiative is not centrally funded and operated. The NNI budget is an aggregation of the nanotechnology-related activities of the participating federal agencies. Congress funds the NNI-related R&D on an agency-by-agency basis, with responsibilities crossing many authorizing committees and appropriations subcommittees. Thus, while it is relatively straightforward to quantify an agency's nanotechnology budget, tracking all NNI investments related to a particular activity—EHS-related research, for example—is much more difficult. The PCAs serve to provide such a tracking mechanism. In addition, according to the 2007 *National Nanotechnology Initiative Strategic Plan*, the division of the Societal Dimensions PCA into two PCAs—Environmental, Health, and Safety; and Educational and Societal Dimensions—is intended to better understand and manage the NNI investment. Such a change indicates a level of flexibility that may enable the executive branch and Congress to more effectively manage and balance investments in discrete areas of the NNI.

NRC recommendation:

> the NSET Subcommittee [should] carry out or commission a study on the feasibility of developing metrics to quantify the return to the U.S. economy from the federal investment in nanotechnology R&D. The study should draw on the Department of Commerce's expertise in economic analysis and its existing ability to poll U.S. industry. Among the activities for which metrics should be developed and relevant data collected are technology transfer and commercial development of nanotechnology.

Few efforts have been made within the federal government to understand the economic impacts of the nation's investments in the NNI. Identification and tracking of data that could serve as an indicator of success in commercializing nanotechnology research or the effects on U.S. job creation or retention has not been formalized. To the extent that federal assessments of the economic contribution of and/or potential for nanotechnology products have occurred, they have not been performed with analytical rigor. Although the Commerce Department retains its economic analysis expertise, resident primarily in the Economics and Statistics Administration's Bureau of Economic Analysis, the Department's Technology Administration, which led Commerce's NNI activities and had government-wide responsibilities for technology transfer activities, was eliminated in August 2007.[82] Prior to its elimination, the Technology Administration contracted for two studies that could contribute to addressing this NRC recommendation: an analysis of barriers to nanotechnology commercialization performed by the University of Illinois at Springfield, and an analysis of innovation metrics conducted by the Alliance for Science and Technology Research in America (ASTRA). These reports are publicly available at Commerce Department websites.[83]

NRC recommendation:

> research on the environmental, health, and safety effects of nanotechnology [should] be expanded. Assessing the effects of engineered nanomaterials on public health and the environment requires that the research conducted be well-defined and reproducible, and that effective methods be developed and applied to (1) estimate the exposure of humans, wildlife, and other ecological receptors to source material; (2) assess effects on human health and ecosystems of both occupational and environmental exposure; and (3) characterize, assess, and manage the risks associated with exposure.

While the NRC asserts the need for additional EHS research, it does not quantify how much more is needed. Clayton Teague, director of the NNCO, has testified that the current level of investment in EHS research is adequate.[84] Many critics from academia, industry, and non-profit organizations have argued strongly that the NNI needs a greater level of investment in EHS research.[85] These critics argue from a variety of perspectives, including the need to:

- protect workers, human health, and the environment;
- create public faith and confidence in the safety of nanotechnology products;
- prevent a problem with one specific nanotechnology product from resulting in a loss of public support for all nanotechnology R&D; and
- create a predictable and stable regulatory environment.

This last factor is deemed by some as critical to fostering future nanotechnology investments.

NRC recommendation:

> the NSET Subcommittee [should] create a working group on education and the workforce that engages the Department of Education and Department of Labor as active participants.

The NSET subcommittee has sought, with limited success, greater involvement of the Departments of Education and Labor in the subcommittee's activities. An NSET subcommittee working group on education and the workforce has not yet been established.

With advocates promising the creation of many new jobs—some assert millions—as a result of global nanotechnology investments, some have expressed concern that the country must prepare students for nanotechnology research, engineering, and production jobs.[86] Assessing which industries are likely to create such jobs, which skills will be needed, and in what timeframe are key challenges. If workers with nanotechnology-specific skills are needed and no workers are available domestically (U.S. citizens, resident aliens, or those in the United States on work visas), potential employers may opt to establish or move operations outside the United States to tap workers with those skills abroad. Conversely, if students are trained for jobs that do not emerge or do not emerge in the same timeframe as students are entering the job market, this investment is lost. In addition, potential students may be discouraged from pursuing future nanotechnology-related studies. Close coordination among the Departments of Commerce, Education, and Labor might help to align federal education and training efforts better with the labor market for nanotechnology workers.

The 21st Century Nanotechnology R&D Act also directed the NRC to address two other issues in its first triennial report: Is molecular self assembly feasible for manufacturing of materials and devices at the nanoscale? And, what are the needs for standards, guidelines, or strategies for ensuring the responsible development of nanotechnology?

Molecular Self-Assembly

Self-assembly is the process by which components (atoms, molecules, or more complex structures) form, without external control or direction, an organized structure. For example, water molecules dispersed in air in cold temperatures can self-assemble to form snowflakes.

Our bodies act as self-assemblers, producing a variety of cells as needed (e.g. to repair a damage to the skin or produce new blood cells from added nutrients).

To what extent can molecular self-assembly be used as a tool for nanomanufacturing? On this issue, the NRC concluded that molecular self-assembly is feasible for the manufacture of simple materials and devices. However, for the manufacture of more sophisticated materials and devices, including complex objects produced in large quantities, the NRC found it unlikely that simple self-assembly processes will yield the desired results. One major barrier cited is the probability of error during assembly as a result of the systems' complexity.

Standards, Guidelines, and Strategies for Ensuring Responsible Development of Nanotechnology

The NRC concluded that it is not possible yet to make a rigorous assessment of the level of environmental and health risks posed by engineered nanomaterials and called for further development of risk assessment protocols. The NRC report also stated that the need for more EHS data requires an expanded research effort to complement dialog on these issues. In addition, until reproducible and well-characterized EHS data are available to inform the development of rigorous risk-based guidelines and best practices, the NRC found it prudent to recommend use of precautionary measures to protect the health and safety of workers, the public, and the environment. The NRC report also stressed that addressing the ethical and societal impacts of nanotechnology will require an integrated approach among scientists, engineers, social scientists, toxicologists, policymakers, and the public.

The National Nanotechnology Initiative at Five Years: Assessment and Recommendations of the National Nanotechnology Advisory Panel, President's Council of Advisors on Science and Technology[87]

This report presents the findings of the first biennial review of the NNI by the President's Council of Advisors on Science and Technology's, acting as the National Nanotechnology Advisory Panel, as mandated by the 21st Century Nanotechnology R&D Act of 2003. The PCAST report finds that the United States is the acknowledged leader in nanotechnology R&D, but the U.S. leadership position is under increasing competitive pressure from growing public and private investments around the world. The report states that the federal investment in the NNI has been well-spent, the United States is well-positioned to maintain global leadership going forward, and continued robust funding is important for long-term U.S. economic well-being and national security. This assessment of the U.S. leadership position is founded not on sales, growth, or market share of commercial products—common measures of global competitiveness for established products—but rather on metrics that may serve as early indicators of potential innovation, such as the U.S. share of scientific publications and patents. The use of such metrics may not be universally accepted as predictive of leadership position. Technological leadership—or even leadership in innovation—does not ensure that the economic benefits from such leadership will accrue to the United States. Companies may choose to manufacture products or conduct other value-added activities outside the United States. If the assessment of national competitiveness is expanded to include the value-added activities and jobs generated or retained within the United States, then the metrics for assessing leadership might change.

The PCAST report acknowledges that there are potential environmental and health risks associated with nanotechnology, but finds that the NNI is directing appropriate attention and adequate resources to the research that will ensure the protection of the public and the environment. Nanotechnology products should not be immune from regulation, according to the report, but such regulation must be rational and based on science, not on perceived fears. The PCAST report states that strong communication exists among the NNI agencies responsible for research and regulation. The PCAST report contains four recommendations for the NNI:

PCAST recommendation:

> To further facilitate technology transfer from the lab to the marketplace, the NNI should expand its interaction with industry, increase federal-state coordination, and improve knowledge management of and access to NNI assets, such as user facilities and instrumentation.

The NSET subcommittee's National Innovation and Liaison with Industry (NILI) working group was established to facilitate NNI interactions with industry, and with state and local nanotechnology initiatives. The NILI working group's limited resources and agency participation have hindered its ability to conduct more extensive and sustained outreach.[88] Due to the structure and resource allocation of the NNI, the initiative's engagements with industry and with state and local initiatives are largely limited to single agency or laboratory interactions and to public engagement activities, such as speeches and information on the NNI website.[89]

PCAST recommendation:

> The NNI should continue its efforts to understand the possible toxicological effects of nanotechnology and where harmful human or environmental effects are proven, pertinent federal agencies should apply appropriate regulatory mechanisms. There should be strong interagency and international collaboration on this issue to eliminate unnecessary duplication of research efforts and to ensure wide dissemination of information. Since exposure to nanomaterials is most likely to occur during the manufacturing process, research on potential hazards associated with workplace exposure must be given the highest priority.

With respect to collaboration on EHS issues, the NSET subcommittee's National Environmental and Health Implications (NEHI) working group is the primary EHS coordination mechanism for participating NNI agencies. The Global Issues in Nanotechnology (GIN) working group works with the NEHI working group on international collaboration on EHS issues. The NIOSH has published several documents addressing concerns about workplace exposure to nanoparticles. *Approaches to Safe Nanotechnology: An Information Exchange With NIOSH* was intended to provide the best currently available knowledge on nanoparticle toxicity and control and to solicit input from the stakeholder community.[90] *Progress Toward Safe Nanotechnology in the Workplace* details the work of NIOSH's Nanotechnology Research Center from 2004 through 2006.[91] In December 2007, NIOSH released interim guidance concerning the medical screening of workers potentially exposed to engineered nanoparticles during the manufacture and industrial use of nanomaterials. The NIOSH says that the document is intended to "generate discussion, fill the current knowledge gap, and provide interim recommendations until further scientific information becomes available."[92] The NIOSH is currently seeking public comment on this guidance.

With respect to regulatory issues associated with nanotechnology, see CRS Report RL34332, *Engineered Nanoscale Materials and Derivative Products: Regulatory Challenges*, by Linda-Jo Schierow.

PCAST recommendation:

> The NNI should establish relationships with the Department of Education and Department of Labor to develop education and training systems to support the Nation's technical proficiency in areas related to nanotechnology.

The PCAST report's recommendation is similar to the recommendation made by the NRC and is discussed earlier in this paper.

PCAST recommendation:

> The NNI must support research aimed at understanding the societal implications of nanotechnology—including ethical, economic and legal implications—and must actively work to inform the public about nanotechnology. The NNI should continue to confront societal issues in an open, straightforward, and science-based manner.

Some critics of the NNI hold deep reservations about the ethical, economic, and legal implications of nanotechnology. Some of these concerns are common to many technologies, such as the allocation of risk and benefit during manufacturing. For example, a neighborhood located near a production facility may bear risks associated with exposure to the byproducts (or products) of manufacturing, while gaining few of the benefits. Concerns about possible adverse effects of nanoscale particles on human health and the environment resulting from their small particle size and unique characteristics may result in increased attention to such costs and benefits with respect to nanoscale material production. Currently, nanotechnology EHS risks are unknown and may be acute or pose no more risk than other manufacturing processes.

Privacy rights are another issue associated with the products of nanotechnology. Nanotechnology may enable the production of highly sensitive, inexpensive sensors that could be deployed ubiquitously in commercial and public settings. While these sensors may allow check-out-free purchases from stores, or monitor the environment for toxic substances, critics argue that they could also impinge on the privacy rights of individuals if, for example, the sensors could detect chemicals related to the use of tobacco, alcohol, or illegal substances without the permission of the individual. Such information might be later applied in law enforcement, life insurance, health insurance, or employment decisions.[93] Others express concern that the economically disadvantaged and less educated—both individuals and nations—might be unable to take part in the benefits that nanotechnology products could offer.[94]

On the legal front, innovations in nanotechnology are already presenting unique challenges to the U.S. Patent and Trademark (USPTO). For example, U.S. case law generally prohibits patenting where the sole element of novelty is a change in size, the characteristic most obviously associated with nanotechnology.[95] In addition, many nanotechnology innovations involve multiple disciplines. Since the USPTO structure for examining patents is discipline-based, an examiner may not have all of the requisite expertise for the examination, affecting both their ability to conduct the examination, and the speed at which it can be done. USPTO also has acknowledged the need to accelerate the speed of nanotechnology-related patent applications. According to John Doll, Commissioner of Patents, the agency is hampered in its ability to recruit and retain patent examiners with the requisite skills to handle nanotechnology patents given the "more generous offers [patent examiners get] from the private sector."[96] Doll said that efforts have been made to improve hiring and retention at USPTO, and that a new processes has been established allowing an accelerated examination of applications.[97]

Selected Nanotechnology Legislation in the 111th Congress

Title I, Subtitle A, H.R. 5116—National Nanotechnology Initiative Amendments Act of 2010

The provisions of Title I, Subtitle A of H.R. 5116, the National Nanotechnology Initiative Amendments Act of 2010, are nearly identical to H.R. 554 (see "H.R. 554—National Nanotechnology Initiative Amendments Act of 2009" below). H.R. 5116 changes the name of the act from the "National Nanotechnology Initiative Amendments Act of 2009," to "National Nanotechnology Initiative Amendments Act of 2010," and removes the term "interdisciplinary" from a provision establishing "green nanotechnology" research centers.

H.R. 554—National Nanotechnology Initiative Amendments Act of 2009

H.R. 554, the National Nanotechnology Initiative Amendments Act of 2009, was introduced on January 15, 2009, and passed by the House of Representatives on February 11, 2009. The bill was referred to the Senate Commerce, Science, and Transportation Committee on February 12, 2009. The purpose of the bill is to authorize activities for support of nanotechnology research and development and for other purposes. Among its provisions, the bill would amend the 21st Century Nanotechnology Research and Development Act of 2003 to:

- require the NSTC triennial strategic plan to include near-term and long-term objectives, the anticipated timeframe for achieving near-term objectives, and metrics for assessing progress; cooperative and collaborative activities in R&D and technology transition supported by the states; and proposed research in areas of national priority;
- require the NSTC annual nanotechnology report supplementing the President's budget request to include a breakout of spending for the development and acquisition of research facilities and instrumentation for each program component area, and a

breakout of spending on all activities related to ethical, legal, environmental, and societal implications;

- direct NNP agencies to support the activities of committees involved in the development of standards for nanotechnology and allow agencies to reimburse the travel costs of scientists and engineers who participate in activities of such committees;
- direct the agencies to fund the National Nanotechnology Coordination Office, and to do so in proportion to each agency's share of the previous year's NNP budget;
- require the NNCO to develop and maintain a publicly accessible database of projects funded under the Environmental, Health, and Safety, the Education and Societal Dimensions, and the Nanomanufacturing program component areas;
- require the NNCO to develop, maintain, and publicize information on nanotechnology facilities supported by the NNP, including at a minimum the terms and conditions for the use of each facility, a description of the capabilities of the instruments and equipment available for use at the facility, and a description of the technical support available to assist users of the facility;
- require the establishment of a National Nanotechnology Advisory Panel (NNAP) "as a distinct entity." Currently, under the provisions of presidential Executive Order 13349, the President's Council of Advisors on Science and Technology serves as the NNAP;[98]
- direct the NNCO to enter into an arrangement with the National Research Council to conduct a triennial review of the NNP, and authorizes funds for FY2010, FY2011, and FY2012; and
- define nanotechnology as "the science and technology that will enable one to understand, measure, manipulate, and manufacture at the nanoscale, aimed at creating materials, devices, and systems with fundamentally new properties or functions," and define nanoscale as "one or more dimensions of between approximately 1 and 100 nanometers."

In addition, the bill would:

- require the designation of a White House Office of Science and Technology Policy associate director to serve as the "Coordinator for Societal Dimensions of Nanotechnology" and would charge the coordinator with convening and chairing a panel of federal agency representatives and others to develop, maintain, implement, and monitor an annual EHS research plan that includes, among other things, standards related to nanotechnology nomenclature; standards for methods and procedures for detecting, measuring, monitoring, sampling, and testing engineered nanoscale materials for environmental, health, and safety impacts; and standard reference materials for EHS testing;
- require the National Science Foundation to provide grants to establish Nanotechnology Education Partnerships to recruit and help prepare secondary school students to pursue postsecondary level courses of instruction in nanotechnology;
- direct the NSTC to establish an Education Working Group under the NSET Subcommittee to coordinate, prioritize, and plan NNP educational activities;

- direct certain NNP agencies to provide companies access to their supported facilities to assist in the development of prototypes of nanoscale products, devices, or processes for determining proof of concept;
- direct NNP agencies to encourage nanotechnology-related submissions to their Small Business Innovation Research (SBIR) and Small Business Technology Transfer (STTR) programs;
- direct NIST to encourage nanotechnology-related submissions to its Technology Innovation Program (TIP), and directs the TIP advisory Board to provide advice to NIST to accomplish this, and to provide an assessment of the adequacy of TIP resources allocated to nanotechnology related projects;
- direct the NSTC to actively pursue industry liaison groups for all industries;
- direct the NNP to coordinate and leverage federal investments with nanotechnology research, development, and technology transition initiatives supported by the States;
- direct the NNP to support nanotechnology R&D activities directed toward application areas that have the potential for significant contributions to national economic competitiveness and for other significant societal benefits, such as nano-electronics, energy efficiency, health care, and water remediation and purification;
- direct the NNP to support research on the development of instrumentation and tools required for the rapid characterization of nanoscale materials and for monitoring of nanoscale manufacturing processes, and to support approaches and techniques for scaling the synthesis of new nanoscale materials to achieve industrial-level production rates; and
- direct certain NNP-supported interdisciplinary research centers to support research on methods and approaches to environmentally benign nanoscale products and nanoscale manufacturing processes, as well as related technology transfer and education activities.

S. 1482—National Nanotechnology Amendments Act of 2009

S. 1482, the National Nanotechnology Amendments Act of 2009, was introduced on July 21, 2009, and referred to the Senate Commerce, Science, and Transportation Committee. The purpose of the bill is to reauthorize the 21st Century Nanotechnology Research and Development Act and to expand the scope of the National Nanotechnology Program (NNP).

Among its provisions, the bill:

- requires the NNP to solicit and draw upon the perspectives of the industrial community to promote the rapid commercial development of nanoscale-enabled devices, systems, and technologies and to coordinate research in determining the key physical and chemical characteristics of nanoparticles and nanomaterials that may pose environmental, health, and safety risks;
- requires the NNCO and other appropriate agencies and councils to issue guidance to agencies that describes a strategy for transitioning research into commercial products and technologies and how the program will coordinate or conduct research on the environmental, health, and safety issues related to nanotechnology;

- requires the NSTC triennial strategic plan to include near-term and long-term objectives, the anticipated timeframe for achieving near-term objectives, and metrics for assessing progress; cooperative and collaborative activities in R&D and technology transition supported by the states; how the NNP intends to encourage and support interdisciplinary research; and proposed research in areas of national priority;
- encourages joint interagency solicitation of grant applications in high priority, multi-disciplinary research areas;
- requires participating agencies to support the activities of the committees of standards setting bodies involved in the development of standards for nanotechnology;
- requires each participating agency to provide funds to support the work of the NNCO. Authorizes appropriations to: (1) NIST for the development of nanotechnology standards; and (2) NSF, for use by the NNCO, to develop and maintain a public information database of NNP projects in EHS; education; public outreach; ethical, legal, and other societal issues; and of nanotechnology facilities accessible for use by individuals from academia and industry;
- makes the National Nanotechnology Advisory Panel (NNAP) a distinct entity, and requires the NNAP to establish a subpanel to enable it to assess whether societal, ethical, legal, environmental, and workforce concerns are adequately addressed by the NNP;
- revises provisions for triennial external review of the NNP;
- requires the designation of a "coordinator for societal dimensions of nanotechnology," within OSTP, to convene a panel to develop a research plan, and requires the coordinator to enter into an arrangement with the National Science Board to create a report that identifies the broad goals and needs of EHS researchers;
- directs the NSTC to establish an interagency Education Working Group to coordinate, prioritize, and plan formal and informal educational activities supported under the NNP, including activities to help participants understand the EHS implications of nanotechnology;
- provides for one or more grants to establish Nanotechnology Education Partnerships to recruit and help prepare secondary school students to pursue postsecondary level courses in nanotechnology;
- requires agencies supporting nanotechnology research facilities to provide access to representatives from industry and other stakeholders for the transfer of research results or assist in developing proof-of-concept prototypes of nanoscale products, devices, or processes;
- directs NIST, in its Technology Innovation Program, and all agencies with Small Business Innovation Research (SBIR) and Small Business Technology Transfer (STTR) programs, to encourage the submission of nanotechnology related grant proposals;
- sets, for the NNP, the objective of establishing industry liaison groups for all industry sectors that would benefit from nanotechnology applications;
- requires coordination and leveraging of federal investments with nanotechnology research, development, and technology transition initiatives supported by state governments;

- requires the NNP to support nanotechnology R&D in areas of national importance (e.g., economic competitiveness, energy production, water purification, agriculture, and health care; in environmental, health, and safety research on the risks of nanoparticles) and in ethical, legal, and societal issues related to nanotechnology;
- requires the NNP to support a wide array of research in support of nanomanufacturing;
- requires the director of the NNCO to review and report on nanomanufacturing research and research facilities;
- requires an NNAP review of the nanomanufacturing program component area and the capabilities of nanotechnology research facilities supported by the NNP;
- sets forth provisions regarding NNP nanoscale characterization and metrology research; and
- requires deliberative public input in the decision making processes affecting policies for the research, development, and use of nanotechnology, and authorizes $2.0 million for the NNCO to carry out this responsibility.

S. 596—Nanotechnology Innovation and Prize Competition Act of 2009

S. 596, the Nanotechnology Innovation and Prize Competition Act of 2009, was introduced on March 16, 2009, and referred to the Senate Commerce, Science, and Transportation Committee. The purpose of the bill is to establish an award program to honor achievements in nanotechnology. Under the bill, the Department of Commerce's National Institute of Standards and Technology is directed to award prizes to individuals and companies for achievement in one or more of the following areas: improvement of the environment, consistent with EPA's Twelve Principles of Green Chemistry; development of alternative energy that has the potential to lessen the dependence of the United States on fossil fuels; and/or improvement of human health, consistent with regulations promulgated by the FDA. The bill authorizes financial prizes for being the first to achieve a specific criteria, as well as recognition prizes, made as part of the previously established National Technology and Innovation Medal award program. The bill authorizes $2,000,000 annually for the financial prizes as well as $750,000 annually for administration of the prize competitions.

H.R. 820—Nanotechnology Advancement and New Opportunities Act

H.R. 820, the Nanotechnology Advancement and New Opportunities Act, was introduced on February 3, 2009, and referred to four House committees: the Committee on Science and Technology, the Committee on Energy and Commerce, the Committee on Ways and Means, and the Committee on Homeland Security. The purpose of the bill is to ensure the development and responsible stewardship of nanotechnology. The bill would:

- establish a $100 million Nanomanufacturing Investment Partnership at the Department of Commerce to work with private investors to advance the commercialization of nanomanufacturing technologies and to increase the commercial application of federally supported research results;

- establish a 15% tax credit, taken over five years, for the purchase of up to $10 million of stock in qualified nanotechnology companies; and
- establish a grant program within the DOC to support the establishment and development of nanotechnology incubators by non-profit entities and degree-granting institutions;
- require the Director of the NNCO to prepare a report to Congress on a nanotechnology research strategy for government and industry that will ensure the development and responsible stewardship of nanotechnology;
- provide a tax credit of 50% for nanotechnology education and training expenses for businesses and individuals;
- authorize an annual appropriation of $15 million for FY2010 through FY2013 for the NSF to conduct a grant program for the development of curriculum materials for interdisciplinary nanotechnology courses at institutions of higher education;
- direct the NSF to establish, through its Advanced Technological Education program, a program to encourage manufacturing companies to enter into partnerships with occupational training centers for the development of training to support nanomanufacturing; and
- direct the Secretary of Energy to submit a report to Congress containing a strategy for increasing interaction among scientists and engineers at DOE national laboratories and the informal science education community to prepare appropriate exhibits for school age children and the general public.

In addition, the bill would amend the 21st Century Nanotechnology Research and Development Act of 2003 to:

- authorize $10 million for the NSF to establish a center for the development of computer-aided design tools for nanotechnology applications;
- authorize an annual appropriation of $30 million for the DOE to conduct a grant program for nanotechnology research to address the need for "clean, cheap, renewable energy;"
- authorize an annual appropriation of $30 million for the EPA for a grant program for nanotechnology research to address technologies for the remediation of pollution and other environmental protection technologies;
- authorize an annual appropriation of $30 million for the DHS to conduct a grant program for nanotechnology research to address the need for sensors and materials related to homeland security needs; and
- authorize an annual appropriation of $30 million for the DHHS to conduct a grant program for nanotechnology research to address health-related applications.

H.R. 2647—National Defense Authorization Act for Fiscal Year 2010

Section 242 of the National Defense Authorization Act for Fiscal Year 2010 (H.R. 2647, P.L. 111-84) amends the Department of Defense's nanotechnology reporting responsibilities to align with those required of other agencies under the 21st Century Nanotechnology Research and Development Act (P.L. 108-153). H.R. 2647 was signed into law on October 28, 2009.

S. 3117—Promote Nanotechnology in Schools Act

S. 3117, the Promote Nanotechnology in Schools Act, was introduced on March 15, 2010, and referred to the Senate Committee on Health, Education, Labor, and Pensions. The purpose of the bill is to strengthen the capacity of eligible institutions (i.e., secondary schools, community colleges, two-year and four-year institutions of higher education, and informal learning science centers) to provide instruction in nanotechnology. The bill would authorize a program at the National Science Foundation for this purpose that would offer eligible institutions grants of up to $400,000 (subject to a 25% match from non-federal sources) to assist in the purchase and maintenance of nanotechnology equipment and software, to develop and provide educational services, and to support teacher education and certification. The bill would authorize $15 million for FY2010 and "such sums as may be necessary" for FY2011 through FY 2013.

H.R. 4502—Nanotechnology Education Act

H.R. 4502, the Nanotechnology Education Act, was introduced on February 19, 2010, and referred to the House Committee on Science and Technology's Subcommittee on Research and Science Education. The purpose of the bill is to strengthen the capacity of eligible institutions (i.e., secondary schools, community colleges, four-year institutions of higher education, and informal learning science centers) to provide instruction in nanotechnology. The bill would authorize a program for this purpose at the National Science Foundation that would offer eligible institutions grants of up to $400,000 (subject to a 25% match from non-federal sources) to assist in the purchase and maintenance of nanotechnology equipment and software, to develop and provide educational services, and to support teacher education and certification. The bill would authorize $40 million for FY2011 and "such sums as may be necessary" for FY2012 through FY 2014.

S. 2942—Nanotechnology Safety Act of 2010

S. 2942, the Nanotechnology Safety Act of 2010, was introduced on January 21, 2010, and referred to the Senate Committee on Health, Education, Labor, and Pensions. The bill would require the Secretary of Health and Human Services to establish within 180 days a program for the scientific investigation of nanoscale materials included or intended for inclusion in FDA-regulated products, to address the potential toxicology of such materials, the effects of such materials on biological systems, and interaction of such materials with biological systems. The bill would authorize $25 million per year for fiscal years 2011 to 2015.

CONCLUDING OBSERVATIONS

Many expect nanotechnology to bring significant economic and societal returns. The United States was the first government to launch a national-level nanotechnology program and has invested more than any other nation. As a result of this focus and these sustained investments, many experts believe that the United States enjoys a technological leadership position in nanotechnology. Other nations are investing heavily and some industrialized and emerging economies have formidable capabilities in nanotechnology. Assessments of the National Nanotechnology Initiative have concluded that the effort is well-managed and has been successful in achieving its objectives so far. However, these assessments have recognized that the NNI faces a variety of challenges in ensuring that the full promise of nanotechnology is realized and that the United States remains the global leader in nanoscale science, engineering, and technology.

Congress may choose to address some or many of the issues addressed in the body of this report in the course of deliberation on the reauthorization of the 21st Century Nanotechnology R&D Act of 2003 or, alternatively, in separate legislation.

The 21st Century Nanotechnology R&D Act's funding authorizations extended through FY2008. Action is being considered in both the House and Senate on reauthorization of the program. Possible topics for consideration in the reauthorization process include budget authorization levels for the covered agencies; R&D funding levels, priorities, and balance across the program component areas; administration and management of the NNI; translation of research results and early-stage technology into commercially viable applications; environmental, health, and safety issues; ethical, legal, and societal implications; education and training for the nanotechnology workforce; metrology, standards, and nomenclature; public understanding; and international dimensions. Consideration may also be given to the establishment of an independent review panel and to coordination of the timing for the NNAP assessment, the NRC assessment, and the NSET subcommittee's strategic plan for the NNI.

APPENDIX A. SELECTED REPORTS ON THE NATIONAL NANOTECHNOLOGY INITIATIVE

Reports of the Nanoscale Science, Engineering, and Technology Subcommittee of the National Science and Technology Council

The National Nanotechnology Initiative: Research and Development Leading to a Revolution in Technology and Industry, Supplement to the President's FY2010 Budget. May 2009.

The National Nanotechnology Initiative: Research and Development Leading to a Revolution in Technology and Industry, Supplement to the President's FY2009 Budget. September 2008.

National Nanotechnology Initiative Fiscal Year 2009 Budget Summary & Highlights. February 2008

Strategy for Nanotechnology-Related Environmental, Health, and Safety Research. February 2008

The National Nanotechnology Initiative Strategic Plan. December 2007.

The National Nanotechnology Initiative: Research and Development Leading to a Revolution in Technology and Industry, Supplement to the President's FY2008 Budget. July 2007.

The National Nanotechnology Initiative: Environmental, Health, and Safety Research Needs for Engineered Nanoscale Materials. September 2006.

The National Nanotechnology Initiative: Research and Development Leading to a Revolution in Technology and Industry, Supplement to the President's FY2007 Budget. July 2006.

The National Nanotechnology Initiative: Research and Development Leading to a Revolution in Technology and Industry, Supplement to the President's FY2006 Budget. March 2005.

The National Nanotechnology Initiative Strategic Plan. December 2004.

Nanotechnology in Space Exploration. August 2004.

Nanoscience Research for Energy Needs. Report from a workshop held in March 2004.

Instrumentation and Metrology for Nanotechnology. Report from a workshop held in January 2004.

Nanotechnology: Societal Implications-Maximizing Benefits for Humanity. Report from a workshop held in December 2003.

Nanobiotechnology. Report from a workshop held in October 2003.

Regional, State, and Local Initiatives in Nanotechnology. Report from a workshop held in September-October 2003.

National Nanotechnology Initiative: Research and Development Supporting the Next Industrial Revolution, Supplement to the President's FY2004 Budget. August 2003.

Nanotechnology and the Environment. Report from a workshop held in May 2003.

National Nanotechnology Initiative: The Initiative and Its Implementation Plan, Detailed Technical Report Associated with the Supplemental Report to the President's FY2003 Budget. June 2002.

National Nanotechnology Initiative: The Initiative and Its Implementation Plan, Detailed Technical Report Associated with the Supplemental Report to the President's FY2001 Budget. July 2000.

Report of the Interagency Working Group on Nanoscience, Technology, and Engineering (NSET Subcommittee Predecessor)

Nanotechnology: Shaping the World Atom by Atom. 1999.

Agency Reports

NIOSH Update: NIOSH Draft Offers Interim Guidance on Medical Screening of Workers Potentially Exposed to Engineered Nanoparticles, National Institute for Occupational Safety and Health. December 2007.

Progress Toward Safe Nanotechnology in the Workplace, National Institute for Occupational Safety and Health. June 2007.

Approaches to Safe Nanotechnology in the Workplace, National Institute for Occupational Safety and Health. July 2006.

Nanoscale Science, Engineering, and Technology in DOE's Office of Basic Energy Sciences, U.S. Department of Energy. February 2003.

External Reviews

A Matter of Size: Triennial Review of the National Nanotechnology Initiative, National Research Council. 2006.

The National Nanotechnology Initiative at Five Years: Assessment and Recommendations of the National Nanotechnology Advisory Panel, President's Council of Advisors on Science and Technology (acting as the National Nanotechnology Advisory Panel). May 2005.

Small Wonders, Endless Frontiers: A Review of the National Nanotechnology Initiative, National Research Council. June 2002.

Appendix B. List of NNI and Nanotechnology-Related Acronyms

ASTRA	Alliance for Science and Technology Research in America
CNST	Center for Nanoscale Science and Technology
CS	Committee on Science
CT	Committee on Technology
CSREES	Cooperative State Research, Education, and Extension Service
DHHS	Department of Health and Human Services
DHS	Department of Homeland Security
DOC	Department of Commerce
DOD	Department of Defense
DOE	Department of Energy
DOJ	Department of Justice
DOT	Department of Transportation
EHS	Environmental, health, and safety
ELSI	Ethical, legal, and societal implications
EPA	Environmental Protection Agency
EOP	Executive Office of the President
EPSCoR	Experimental Program to Stimulate Competitive Research

FHWA	Federal Highway Administration
GIN	Global Issues in Nanotechnology working group
ISO	International Standards Organization
IWGN	Interagency Working Group on Nanotechnology
NASA	National Aeronautics and Space Administration
NCI	National Cancer Institute
NEHI	National Environmental and Health Implications working group
NGO	Non-governmental organization
NIH	National Institutes of Health
NILI	National Innovation and Liaison with Industry working group
NIOSH	National Institute of Occupational Safety and Health
NIST	National Institute of Standards and Technology
NNAP	National Nanotechnology Advisory Panel
NNCO	National Nanotechnology Coordination Office
NNI	National Nanotechnology Initiative
NNIN	National Nanotechnology Infrastructure Network
NNN	National Nanomanufacturing Network
NNP	National Nanotechnology Program
NPEC	Nanotechnology Public Engagement and Communications working group
NRC	National Research Council
NSET	Nanoscale Science, Engineering, and Technology subcommittee
NSF	National Science Foundation
NSEC	Nanoscale Science and Engineering Center
NSRC	Nanoscale Science Research Centers
NSTC	National Science and Technology Council
OECD	Organization for Economic Cooperation and Development
OMB	Office of Management and Budget
OSTP	Office of Science and Technology Policy
PCA	Program Component Areas
PCAST	President's Council of Advisors on Science and Technology
R&D	Research and development
SBIR	Small Business Innovation Research
STTR	Small Business Technology Transfer Research
TSA	Transportation Security Administration
USDA	U.S. Department of Agriculture
USPTO	U.S. Patent and Trademark Office

End Notes

1 President Obama requested $1.6 billion in nanotechnology R&D funding for FY2010. Although Congress has passed all of the regular appropriations acts, the precise amount provided for FY2010 is currently under evaluation by the agencies and the White House Office of Management and Budget and will likely be published with the President's FY2011 budget request in February 2010.

2 In the context of the NNI and nanotechnology, the nanoscale refers to a dimension of 1 to 100 nanometers (see box on this page).

[3] While extensive R&D has been, and continues to be, conducted to understand and harness the properties of individual atoms, this is not the domain of nanotechnology.

[4] *Nanotechnology Takes a Deep Breath...and Prepares to Save the World,* Cientifica, April 2009, available at http://www.cientifica.eu/files/Whitepapers/Nanotechnology%20Takes%20a%20Deep%20Breath.pdf.

[5] The National Nanotechnology Initiative at Five Years: Assessment and Recommendations of the National Nanotechnology Advisory Panel, President's Council of Advisors on Science and Technology, May 2005, available at http://www.nano.gov/html/res/FINAL_PCAST_NANO_REPORT.pdf.

[6] *Nanotechnology Takes a Deep Breath...and Prepares to Save the World,* Cientifica, April 2009, available at http://www.cientifica.eu/files/Whitepapers/Nanotechnology%20Takes%20a%20Deep%20Breath.pdf.

[7] Ibid.

[8] "Overhyped Technology Starts to Reach Potential," press release, Lux Research, July 22, 2008. http://www.luxresearchinc.com/press/RELEASE_Nano-SMR_7_22_08.pdf. Numbers do not add to total due to rounding.

[9] Ibid

[10] *Halfway to the Trillion Dollar Market: A Critical Review of the Diffusion of Nanotechnologies*, Cientifica, 2007. http://www.cientifica.eu/files/Whitepapers/A%20Reassessment%20of%20the%20Trillion%20WP.pdf

[11] While views vary on how to calculate nanotechnology's contribution to these products, the consensus is that nanotechnology is likely to have a significant economic impact and transformative effect on many industries.

[12] National Cancer Institute website. http://nano.cancer.gov/resource_center/tech_backgrounder.asp

[13] Ibid.

[14] Aslan, Kadir; Lakowicz, Joseph R.; and Geddes, Chris D. "Nanogold plasmon resonance-based glucose sensing. Wavelength-ratiometric resonance light scattering," *Analytical Chemistry*, 2005, Vol. 77. *Strategic Plan for Pediatric Urology,* National Institute of Diabetes and Digestive and Kidney Disease, National Institutes of Health, Department of Health and Human Services, February 2006.

[15] Risbud, Aditi. "Fruit of the Nano Loom," *Technology Review*, February 2006.

[16] *Nanoscience Research for Energy Needs,* Nanoscale Science, Engineering, and Technology Subcommittee, National Science and Technology Council, The White House, December 2004.

[17] Risbud, Aditi. "Cheap Drinking Water from the Ocean," *Technology Review*, June 2006.

[18] Selko, Adrienne. "New Nanotechnology-Based Coatings Are Energy Efficient and Environmentally Sound," *Industry Week*, August 22, 2007. "Tomorrow's Green Nanofactories," *Science Daily*, July 11, 2007.

[19] *National Nanotechnology Initiative—Leading to the Next Industrial Revolution,* Interagency Working Group on Nanoscience, Engineering, and Technology, National Science and Technology Council, The White House. http://www.ostp.gov/NSTC/html/iwgn/iwgn.fy01budsuppl/nni.pdf

[20] *21st Century Agriculture: A Critical Role for Science and Technology*, U.S. Department of Agriculture, June 2003; and *Nanoscale Science and Engineering for Agriculture and FoodSystems: Draft Report of the National Planning Workshop to the Cooperative State Research, Education, and Extension Service of the U.S. Department of Agriculture*, July 2003.

[21] *Nanotechnology in Space Exploration*, Nanoscale Science, Engineering, and Technology Subcommittee, National Science and Technology Council, The White House, August 2004, available at http://www.nano.gov/nni_space_exploration_rpt.pdf.

[22] *Nanotechnology and the Environment*, Nanoscale Science, Engineering, and Technology Subcommittee, National Science and Technology Council, The White House, May 2003, available at http://www.nano.gov/NNI_Nanotechnology_and_the_Environment.pdf.

[23] *Proceedings of the U.S. Environmental Protection Agency Workshop on Nanotechnology for Site Remediation,* U.S. Environmental Protection Agency, October 2005.

[24] "National Nanotechnology Initiative: Leading to the Next Industrial Revolution," press release, The White House, January 21, 2000. http://clinton4.nara.gov/WH/New/html/20000121_4.html; and "Steering the technology that will redefine life as we know it," *Industrial Biotechnology*, Vol. 1, No. 3, Fall 2005. http://www.nsf.gov/crssprgm/nano/ reports/mcr_ind_biotech_interview.pdf

[25] This figure includes $1.657 billion in regular appropriations and $0.140 billion in funding provided by the American Recovery and Reinvestment Act of 2009 (P.L. 111-5).

[26] While many provisions of this act have no sunset provision, FY2008 was the last year of agency authorizations included in the act.

[27] Prior to P.L. 108-153, the Bob Stump Defense Authorization Act for Fiscal Year 2003 (P.L. 107-314) required DOD to "provide for interagency cooperation and collaboration on nanoscale research and development." The NSET subcommittee is a subcommittee of the NSTC Committee on Technology.

[28] Metrology is the science of measurement, including the equipment and processes used to produce a measurement.

[29] The Department of Commerce has characterized emerging and enabling technologies as those that "offer a wide breadth of potential application and form an important technical basis for future commercial applications." (ATP Rule, 15 C.F.R. Part 295).

[30] The term "valley of death" is used by business executives, economists, and venture capitalists to describe the development gap that often exists between a laboratory discovery and the market's willingness to invest to

advance the discovery to a final commercial product. This gap occurs due to a variety of issues, such as technical risk, market uncertainty, and likelihood of obtaining an adequate return on investment.

[31] Crews, Clyde Wayne, Jr., "Washington's Big Little Pork Barrel: Nanotechnology," Cato Institute website, May 29, 2003.

[32] Testimony of Andrew Maynard, Chief Science Advisor, Project on Emerging Nanotechnologies, Woodrow Wilson International Center for Scholars, "Research on Environmental and Safety Impacts of Nanotechnology: Current Status of Planning and Implementation under the National Nanotechnology Initiative," hearing, Subcommittee on Research and Science Education, House Committee on Science and Technology, October 31, 2007.

[33] "NGOs urge precautionary principle in use of nanomaterials," EurActiv.com, June 14, 2007. http://www.euractiv.com/en/environment/ngos-urge-precautionary-principle-use-nanomaterials/article-164619 Sass, Jennifer. "Nanotechnology and the Precautionary Principle," presentation, Natural Resources Defense Council, 2006. http://docs.nrdc.org/health/hea_06121402a.pdf The precautionary principle has been used in other countries on some issues. For example, the Biosafety Protocol to the 1992 Convention on Biological Diversity incorporates provisions applying the precautionary principle to the safe handling, transfer, and trade of genetically modified organisms. For further information, see CRS Report RL30594, *Biosafety Protocol for Genetically Modified Organisms: Overview*, by Alejandro E. Segarra and Susan R. Fletcher.

[34] "No Small Matter II: The Case for a Global Moratorium—Size Matters!," Occasional Paper Series, ETC Group, April 2003. http://www.etcgroup.org/upload/publication/pdf_file/165

[35] *The National Nanotechnology Initiative Strategic Plan,* Nanoscale Science, Engineering, and Technology Subcommittee, National Science and Technology Council, The White House, December 2004, available at http://www.nano.gov/NNI_Strategic_Plan_2004.pdf.

[36] Ibid.

[37] National Nanotechnology Initiative website. http://www.nano.gov/html/about/history.html

[38] "National Nanotechnology Initiative: Leading to the Next Industrial Revolution," press release, The White House, January 21, 2000. http://clinton4.nara.gov/WH/New/html/20000121_4.html; and National Nanotechnology Initiative website. http://www.nano.gov/html/about/history.html

[39] In its January 21, 2001 press release, "National Nanotechnology Initiative: Leading to the Next Industrial Revolution," announcing the establishment of the NNI, the White House identified only six participating agencies—NSF, DOD, DOE, NIST, NASA, and NIH. Subsequently, EPA and DOJ reported nanotechnology R&D funding in FY2001, bringing the total number of agencies funding nanotechnology R&D in FY2001 to eight.

[40] NNI participants include agencies that either conduct or provide funding for nanotechnology R&D, as well as agencies with missions that may affect the development, commercialization, and use of nanotechnology. For example, in the latter case, the Food and Drug Administration may regulate (or not regulate) nanotechnology products, the U.S. Patent and Trademark Office's (USPTO) treatment of nanotechnology-related patents may affect the value of the underlying intellectual property, and the execution of the missions of the Departments of Education and Labor could affect the preparedness of the U.S. workforce for emerging nanotechnology jobs. Some nanotechnology R&D agencies may also have non-R&D missions related to nanotechnology. For example, EPA conducts and funds R&D but also has a regulatory mission that could affect nanotechnology research, development, production, use, and/or disposal.

[41] The original six agencies identified at the launch of the NNI were the Department of Defense, Department of Energy, National Institute of Standards and Technology (Department of Commerce), National Science Foundation, National Aeronautics and Space Administration, and National Institutes of Health (DHHS). "National Nanotechnology Initiative: Leading to the Next Industrial Revolution," press release, The White House, January 21, 2000. http://clinton4.nara.gov/WH/New/html/20000121_4.html; and National Nanotechnology Initiative website. http://www.nano.gov/html/about/history.html

[42] U.S. Congress. 2003. 21st Century Nanotechnology Research and Development Act. P.L. 108-153. 15 U.S.C. 7501. 108 Cong., December 3.

[43] FY2003 funding attributed to DHS for the purpose of this calculation is based on nanotechnology R&D appropriations received by the Department of Transportation's Transportation Security Administration (TSA). TSA was transferred to DHS in the Homeland Security Act of 2002 (P.L. 107-296) which was enacted after the start of FY2003.

[44] *The National Nanotechnology Initiative Strategic Plan*, Nanoscale Science, Engineering, and Technology Subcommittee, National Science and Technology Council, The White House, December 2004, available at http://www.nano.gov/NNI_Strategic_Plan_2004.pdf.

[45] The agencies that participate in the NSET subcommittee comprise the NNI. NSET subcommittee members include Bureau of Industry and Security, Department of Commerce; Consumer Product Safety Commission; National Institute of Food and Agriculture, Department of Agriculture; Department of Defense; Department of Education; Department of Energy; Department of Homeland Security; Department of Justice; Department of Labor; Department of State; Department of Transportation; Department of the Treasury; Environmental Protection Agency; Food and Drug Administration; Forest Service, Department of Agriculture; Director of

National Intelligence; International Trade Commission; National Aeronautics and Space Administration; National Institutes of Health, U.S. Department of Health and Human Services; National Institute for Occupational Safety and Health, Center for Disease Control, Department of Health and Human Services; National Institute of Standards and Technology, Department of Commerce; National Science Foundation; Nuclear Regulatory Commission; U.S. Geological Survey; and U.S. Patent and Trademark Office, Department of Commerce. The Department of Commerce's Technology Administration was a participating agency in the NNI until its elimination in August 2007 under the America COMPETES Act (P.L. 110-69).

[46] The National Nanotechnology Initiative: Research and Development Leading to a Revolution in Technology and Industry-Supplement to the President's FY2008 Budget, Nanoscale Science, Engineering, and Technology Subcommittee, National Science and Technology Council, The White House, July 2007, available at http://www.nano.gov/NNI_08Budget.pdf.

[47] Testimony of E. Clayton Teague, Director, National Nanotechnology Coordinating Office, *Research on Environmental and Safety Impacts of Nanotechnology: Current Status of Planning and Implementation under the National Nanotechnology Initiative*, hearing, Subcommittee on Research and Science Education, House Committee on Science and Technology, October 31, 2007.

[48] The Industrial Research Institute is an association of companies and federally funded laboratories with the mission of improving R&D capabilities through the development and dissemination of best practices.

[49] National Nanotechnology Initiative website, http://www.nano.gov.

[50] The eight agencies that are signatories to the memorandum of understanding are NSF, DOD, DOE, NIH, NIST, NASA, EPA, and DOT.

[51] This figure does not include NNI funding provided by ARRA. Agency ARRA allocations included $511.2 million for nanotechnology R&D. Agencies may report additional ARRA funding for SBIR and STTR projects later, when 2009 SBIR/STTR data become available.

[52] *The National Nanotechnology Initiative Strategic Plan,* Nanoscale Science, Engineering, and Technology Subcommittee, National Science and Technology Council, The White House, December 2004, http://www.nano.gov/ NNI_Strategic_Plan_2004.pdf.

[53] *The National Nanotechnology Initiative Strategic Plan,* Nanoscale Science, Engineering, and Technology Subcommittee, National Science and Technology Council, The White House, December 2007, available at http://www.nano.gov/NNI_Strategic_Plan_2007.pdf.

[54] *The National Nanotechnology Initiative: FY2010 Budget & Highlights*, Nanoscale Science, Engineering, and Technology Subcommittee, National Science and Technology Council, The White House, May 2009, available at http://www.nano.gov/NNI_2010_budget_supplement.pdf.

[55] Ibid.

[56] Ibid.

[57] Ibid.

[58] Top-down processes are those that achieve design features by removing material from a larger block of material; bottom-up processes begin with smaller building blocks (atoms or molecules) and achieve design features by putting them together, possibly using self-assembly.

[59] *The National Nanotechnology Initiative: FY2010 Budget & Highlights*, Nanoscale Science, Engineering, and Technology Subcommittee, National Science and Technology Council, The White House, May 2009, available at http://www.nano.gov/NNI_2010_budget_supplement.pdf.

[60] Ibid.

[61] *The National Nanotechnology Initiative: FY2010 Budget & Highlights*, Nanoscale Science, Engineering, and Technology Subcommittee, National Science and Technology Council, The White House, May 2009, available at http://www.nano.gov/NNI_2010_budget_supplement.pdf.

[62] Ibid.

[63] *The National Nanotechnology Initiative: FY2010 Budget & Highlights*, Nanoscale Science, Engineering, and Technology Subcommittee, National Science and Technology Council, The White House, May 2009, available at http://www.nano.gov/NNI_2010_budget_supplement.pdf.

[64] Ibid.

[65] In addition, 18 other Materials Research Science and Engineering Centers conduct nanotechnology-related research as part of their overall efforts.

[66] Private telephone communication between CRS and NSTC staff, January 31, 2008.

[67] Private e-mail communication between CRS and NSF staff, January 31, 2008.

[68] Private telephone communication between CRS and NSTC staff, January 31, 2008.

[69] Private e-mail communication between CRS and NSF staff, January 31, 2008.

[70] *The National Nanotechnology Initiative: FY2011 Budget & Highlights*, Nanoscale Science, Engineering, and Technology Subcommittee, National Science and Technology Council, The White House, February 2010, available at http://www.nano.gov/NNI_2011_budget_supplement.pdf.

[71] *The National Nanotechnology Strategic Plan*, Nanoscale Science, Engineering, and Technology Subcommittee, National Science and Technology Council, The White House, December 2007, available at http://www.nano.gov/ NNI_Strategic_Plan_2007.pdf.

[72] *Strategy for Nanotechnology-related Environmental, Health, and Safety Research*, Science, Engineering, and Technology Subcommittee, National Science and Technology Council, The White House, February 2008, available at http://www.nano.gov/NNI_EHS_Research_Strategy.pdf.

[73] *The National Nanotechnology Initiative: Environmental, Health, and Safety Research Needs for Engineered Nanoscale Materials*, Nanoscale Science, Engineering, and Technology Subcommittee, National Science and Technology Council, The White House, September 2006, available at http://www.nano.gov/NNI_EHS_research_needs.pdf.

[74] Prioritization of Environmental, Health, and Safety Research Needs for Engineered Nanoscale Materials: An Interim Document for Public Comment, Nanoscale Science, Engineering, and Technology Subcommittee, National Science and Technology Council, The White House, August 2007, available at http://www.nano.gov/Prioritization_EHS_Research_Needs_Engineered_Nanoscale_Materials.pdf.

[75] *The National Nanotechnology Initiative: Environmental, Health, and Safety Research Needs for Engineered Nanoscale Materials*, Nanoscale Science, Engineering, and Technology Subcommittee, National Science and Technology Council, The White House, September 2006, available at http://www.nano.gov/NNI_EHS_research_needs.pdf.

[76] The National Research Council, the National Academy of Sciences, the National Academy of Engineering, and the Institute of Medicine are part of a private, non-profit institution established under a congressional charter. They are collectively referred to as the National Academies.

[77] P.L. 108-153 directs the President to "establish or designate a National Nanotechnology Advisory Panel." In July 2004, President Bush designated the President's Council of Advisors on Science and Technology to serve as the NNAP by issuing Executive Order 13349, *Amending Executive Order 13226 To Designate the President's Council of Advisors on Science and Technology To Serve as the National Nanotechnology Advisory Panel.*

[78] The National Nanotechnology Initiative: Second Assessment and Recommendations of the National Nanotechnology Advisory Panel, President's Council of Advisors on Science and Technology, May 2005, available at http://www.whitehouse.gov/sites/default/files/microsites/ostp/PCAST-NNAP-NNI-Assessment-2008.pdf.

[79] *Addendum to The National Nanotechnology Initiative: Second Assessment and Recommendations of the National Nanotechnology Advisory Panel*, President's Council of Advisors on Science and Technology, May 2005, available at National Research Council, 2006, available at http://www.whitehouse.gov/sites/default/files/microsites/ostp/PCAST-Addendum-Letter.pdf.

[80] Ibid.

[81] *A Matter of Size: Triennial Review of the National Nanotechnology Initiative*, National Research Council, 2006, available at http://books.nap.edu/catalog.php?record_id=11752.

[82] The Technology Administration was eliminated in the America COMPETES Act (P.L. 110-69).

[83] *Barriers to Nanotechnology Commercialization*, College of Business Management, University of Illinois at Springfield, September 2007, available at http://www.osec.doc.gov/Report-Barriers%20to%20Nanotechnology%20Commercialization.pdf *Innovation Vital Signs Project*, Alliance for Science and Technology Research in America, July 2007. http://www.ntis.gov/ta_reports/Report-InnovationVitalSigns.pdf.

[84] Testimony of Clayton Teague, director of the NNCO, hearing, "Environmental and Safety Impacts of Nanotechnology: What Research is Needed?" House Committee on Science, November 17, 2005. http://commdocs.house.gov/committees/science/hsy24464.000/hsy24464_0.HTM

[85] Ibid.

[86] Phillip J. Bond, Under Secretary for Technology, U.S. Department of Commerce, remarks, "Nanotechnology: Economic Opportunities, Societal and Ethical Challenges," NanoCommerce 2003, December 9, 2003. http://www.technology.gov/Speeches/PJB_031209.htm *Sizing Nanotechnology's Value Chain*, Lux Research, October 2004.

[87] The National Nanotechnology Initiative at Five Years: Assessment and Recommendations of the National Nanotechnology Advisory Panel, President's Council of Advisors on Science and Technology, May 2005, available at http://www.nano.gov/html/res/FINAL_PCAST_NANO_REPORT.pdf.

[88] Private telephone and e-mail communication with Sean Murdock, executive director of the NanoBusiness Alliance, February 4, 2008.

[89] Ibid.

[90] *Approaches to Safe Nanotechnology: An Information Exchange with NIOSH*, National Institute for Occupational Safety and Health, July 2006.

[91] *Progress Toward Safe Nanotechnology in the Workplace*, National Institute for Occupational Safety and Health, June 2007.

[92] *NIOSH Update: NIOSH Draft Offers Interim Guidance on Medical Screening of Workers Potentially Exposed to Engineered Nanoparticles*, National Institute for Occupational Safety and Health, December 13, 2007.

[93] Moore, Fiona M., "Implications of Nanotechnology Applications: Using Genetics as a Lesson," *Health Law Review*, Vol. 10, No. 3, 2002. http://www.law.ualberta.ca/centres/hli/pdfs/hlr/v10_3/10.3moorefrm.pdf

[94] Smith, Richard H.,"Social, Ethical, and Legal Implications of Nanotechnology," *Societal Implications of Nanoscience and Nanotechnology* (The Netherlands:Kluwer Academic Publishers, 2001).

[95] *Nanotechnology Patents: Issues for Nanotechnology Inventions*, Dorsey and Whitney, LLP, May 9, 2005.
[96] Doll, John, Commissioner of Patents, U.S. Patent and Trademark Office, Letter to the Editor, *Small Times*, April 23, 2007. http://www.smalltimes.com/Articles/Article_Display.cfm?ARTICLE_ID=290818&p=109
[97] Ibid.
[98] Executive Order 13349 is available at http://edocket.access.gpo.gov/cfr_2005/janqtr/3CFR13349.htm.

INDEX

D

S